SpringerBriefs in Applied Sciences and Technology

Series Editor

Andreas Öchsner, Griffith School of Engineering, Griffith University, Southport, QLD, Australia

SpringerBriefs present concise summaries of cutting-edge research and practical applications across a wide spectrum of fields. Featuring compact volumes of 50 to 125 pages, the series covers a range of content from professional to academic. Typical publications can be:

- A timely report of state-of-the art methods
- An introduction to or a manual for the application of mathematical or computer techniques
- A bridge between new research results, as published in journal articles
- A snapshot of a hot or emerging topic
- An in-depth case study
- A presentation of core concepts that students must understand in order to make independent contributions

SpringerBriefs are characterized by fast, global electronic dissemination, standard publishing contracts, standardized manuscript preparation and formatting guidelines, and expedited production schedules.

On the one hand, **SpringerBriefs in Applied Sciences and Technology** are devoted to the publication of fundamentals and applications within the different classical engineering disciplines as well as in interdisciplinary fields that recently emerged between these areas. On the other hand, as the boundary separating fundamental research and applied technology is more and more dissolving, this series is particularly open to trans-disciplinary topics between fundamental science and engineering.

Indexed by EI-Compendex, SCOPUS and Springerlink.

More information about this series at http://www.springer.com/series/8884

Ramashini Murugaiya
Manisha Milani Mahagammulle Gamage
Krishani Murugiah • Madhumathy Perumal

Acoustic-Based Applications for Vertebrate Vocalization

Springer

Ramashini Murugaiya (iD)
Department of Computer Science and
Informatics
Uva Wellassa University
Badulla, Sri Lanka

Krishani Murugiah (iD)
Department of Biotechnology
Pavendar Bharathidasan College of
Engineering
Trichirapalli, Tamil Nadu, India

Manisha Milani Mahagammulle Gamage (iD)
Faculty of Integrated Technologies
Universiti Brunei Darussalam
Bandar Seri Begawan, Brunei Darussalam

Madhumathy Perumal (iD)
RV Institute of Technology
and Management
Bangalore, India

ISSN 2191-530X   ISSN 2191-5318   (electronic)
SpringerBriefs in Applied Sciences and Technology
ISBN 978-3-030-85772-1   ISBN 978-3-030-85773-8   (eBook)
https://doi.org/10.1007/978-3-030-85773-8

This Springer imprint is published by the registered company Springer Nature Switzerland AG
The registered company address is: Gewerbestrasse 11, 6330 Cham, Switzerland

Preface

This book shows a pool of potential applications that can be invented using vertebrate vocalisation along with the collaboration of modern technologies. The overall aim of this book is to provide a foundation for future researches to build upon by providing knowledge for identifying the gap between the trending technology and nature to address a vast range of issues for ecology-related application, conservation-based projects, and entertainment-based applications. Moreover, this may open up new paths for identifying new sounds as well as species in nature which will be the essential tool for the sustainability of the ecosystem.

Chapter 1 emphasises the importance of the vertebrate vocalization-related applications through elaborating various classes of vertebrates and their vocal production with appropriate examples. Additionally, the primary factors that involve in the development, along with the challenges faced during implementation, will be discussed in depth. Then the most recent works related to vertebrate vocals shall be critically reviewed. Chapter 2 includes the significance of feature extraction in vertebrate vocalisation. However, the importance of feature engineering and feature selection techniques will be discussed along with some feature analysis experiments using statistical tests. In Chap. 3, the future directions of vertebrate vocalisation that can be incorporated with trending technologies will be provided.

Overall, this book can be used as a guide by budding researchers while starting their career path in vertebrate vocal and natural sound-related research to solve an enormous range of domains such as ecology, tourism, entertainment, conservation, education, and biology.

Badulla, Sri Lanka   Murugaiya Ramashini
Bandar Seri Begawan, Brunei   M. G. Manisha Milani
Trichy, India   Murugiah Krishani

Contents

Chapter 1
Introduction to Applications on Vertebrate Vocalisation

1.1 Acoustics of Ecosystem

Sounds in nature are like an icing on the cake, which adds more beauty to nature such as the soothing sound of river streaming over the landscape, gushing waterfall and delightful chirping of birds. For many biotic factors, the sound is art for communicating information about their surroundings. Disturbance in sound may lead to fall in grazing efficiency, least mating opportunities and even declined survival rates. Natural sound especially animal sound and birds chirping acts as a miner's canary which warns the coming of great danger such as climatic change, excessive pollution and other environmental alterations. On the whole, the ultimate goal of the audio-based applications on the acoustics of living creatures is to conserve the environment.

In 1969, South worth, an urban planner, used the term "soundscape" which means sounds occurring over an area. Based on the source, the soundscape is classified in three ways, namely, biophony, geophony and anthrophony. Geographical sound includes thunder, rain, movement of water and windfalls under the geophony. The sounds produced by human-made activities such as machinery and vehicles are listed under anthrophony, and the third one is the biophony which we mainly discussed in this book that is the collection of sound produced by animals [1].

1.2 Significance of Vocalisation

The animal sound was pivotal. Thus animals produce sound for expressing their feelings, including wooing, mating, parent-offspring interactions and even to put up a fight against their territory. Through the animals having various ability to produce sounds like stridulation, flapping wings, vocalisation is considered as quite

© The Author(s), under exclusive license to Springer Nature
Switzerland AG 2022
R. Murugaiya et al., *Acoustic-Based Applications for Vertebrate Vocalization*,
SpringerBriefs in Applied Sciences and Technology,
https://doi.org/10.1007/978-3-030-85773-8_1

dominant. It is the best way of expressing their emotions and needs such as hunger, thirst and appetite [2]. Furthermore, another necessity is to prevent them from endangered species by capturing their vocal sounds when they are in distress and danger. Therefore considering these significants in this book, we discussed the applications related to the vocalisation of vertebrates.

1.2.1 Vocalisation: An Innate Existence of Vertebrates

Based on physics, the molecules get vibrated, and travel in the form of audible waves in a medium like a gas, a solid or a liquid is termed as sound, which depicts sound cannot occur in a vacuum or outer space. Hence sound generation in animals entirely relied on their respiration patterns. The vocal folds in the larynx produce sound for mammals, and the membranes in the syrinx of the birds help to vibrate the tissues for producing sound. The nervous system consists of the brainstem, and spinal cord resides a neuronal circuit which directs the respiratory organs to undergo vocal communication. Thus vocalisation is tightly coupled with neuronal circuits, and so it is considered as a congenital behaviour of vertebrates [3]. Animals with a spinal cord surrounded by cartilage or bone are termed as vertebrates, that is, amphibians, fishes, reptiles, birds and mammals are listed under vertebrates [4]. The classification and characteristic features of vertebrates are shown in Fig. 1.1.

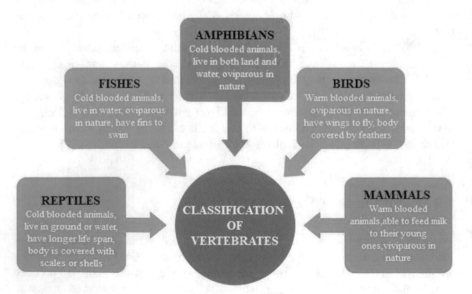

Fig. 1.1 Classification of vertebrates and its characteristic features

1.3   Vocal Mechanism in Different Vertebrates

In mammals, there is a more significant divergence in the frequency of the vocal sound produced among different species. For instance, rodents produce high frequency or ultrasounds above 20 kHz, whilst the elephants produce low frequency or infrasounds under 20 Hz. Mammals use the larynx to produce sound. The larynx in humans consists of two vocal folds and four muscles interconnected by cartilages, namely, interarytenoid, thyroarytenoid, cricothyroid and posterior cricoarytenoid muscles. Opening and closing of glottis, an aperture between vocal folds, plays a vital role in vocalisation and basal breathing. Air gets moved into larynx throughout basal breathing in which posterior cricoarytenoid muscles get contracted and pulls the vocal folds and arytenoid cartilages away from each other, whilst in the course of vocalisation, interarytenoid muscles get contracted to unite the vocal folds for closing the glottis and acts as a barrier for airflow. Exhaled air creates an airflow pressure that collides with vocal folds and causes vibrations to produce sound [3]. The anatomical structure of organs which are involved in human vocalisation is shown in Fig. 1.2.

The vocal organ for birds is syrinx and is located in the air sac of the chest, and it is also connected to another air sac and lungs. The syrinx of a few aves such as pigeons, doves and parrots is located at the lower end of the trachea. This is shown in Fig. 1.3, in the form of modified cartilage rings. Birds like owls, penguins and some cuckoos, the syrinx exists as a pair of partial syrinxes found in the primary bronchi. However, in most of the birds, syrinx is located at the centre of two primary bronchi that joins the trachea. Syrinx consists of strong double-barrelled muscles commonly known as syringeal muscles, cartilages, a vocal tract and labia, a vibrating membrane. Like mammals and amphibians, expiratory airflow produces vocal

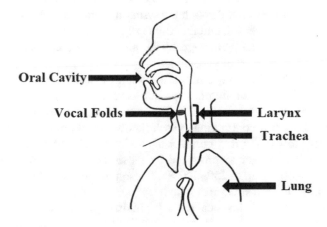

Fig. 1.2 Anatomy of vocal organs in the human

Fig. 1.3 Anatomy of vocal
organs in birds

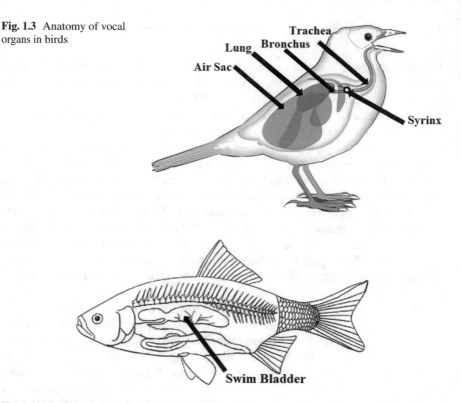

Fig. 1.4 Position of the swim bladder in the Bony fish

sounds in birds, but it is very efficient compared to others. During expiration, the muscle contracts the abdomen and thorax by increasing the respiratory pressure of the air sac and then flows out through the syrinx and trachea, whilst in the inspiration, the process gets reversed, and the respiratory pressure gets decreased in the air sac. Then vocal sounds are generated by the airflow that induces the vibration of elements present in the wall of syrinx [3, 5].

Marine vertebrates like bony fishes use anatomical structures like swim bladder, pectoral muscles or head structures to produce sound. A gas-filled organ known as a swim bladder is depicted in Fig. 1.4. The swim bladder is used to control buoyancy, which has a fast-contracting muscle known as drumming muscles which vibrate actively to expand and contract the swim bladder to generate drumming sounds. Generally, the sound produced by these methods is having frequencies ranging from 45–60 Hz up to 250–300 Hz. It also produces vocalisation by rubbing pectoral muscles or head structures with body parts [3].

Along with the larynx, the vocal sac is the main organ responsible for producing sound in the amphibians. Vocal sac is the specialised anatomical structure in amphibians which is inflated with air and helps the species to survive under the water when the mouth and nostrils remain close. The vibration of tissues produces the vocal sounds of frogs during expiration. The air in the lungs gets expelled out to the larynx, as explained in Fig. 1.5, and causes vibrations in itself and the vocal sac, and then it enters into the buccal cavity and passes through the vocal slits to expand the vocal sac. Vocal sac acts as a resonance chamber to produce vocalisation [3, 6].

Hissing is a distinctive sound produced by reptiles as a threat display. In snakes, the movement of muscles attached to the ribs causes lung ventilation resulting in increased pressure of the posterior air sac. The excessive pressure gets forcibly expelled from the air sac to the larynx through the trachea and finally reaches the oral cavity. Snakes have an aperture in the oral cavity called the glottis. Due to the expulsion of excess air, the glottis gets rattled to produce hiss sound [7]. Lizards and crocodiles 'hiss' during predation, interactions with other individuals and defensive behaviour. Some turtles and crocodiles are hissing or roaring when they feel excited. The anatomy of vocal organs in lizard is shown in Fig. 1.6.

The calls of crocodile for breeding and gecko calls produced by some lizards are claims under reptile vocalisation. These kinds of sounds are used for distress, courtship and territorial disputes. A study on the mechanism of sound production on lizard reported that the mouth of the lizard is opened during the call, and this call always appears during expiration. The lungs get curtailed to expel air, and then the anterior region of the arytenoid moves laterally to place the vocal cord perpendicular to the animal's axis. It tensed and collided with each other, and the glottis shown in Fig. 1.6 remains open. The vibration of the vocal cord due to airflow results in

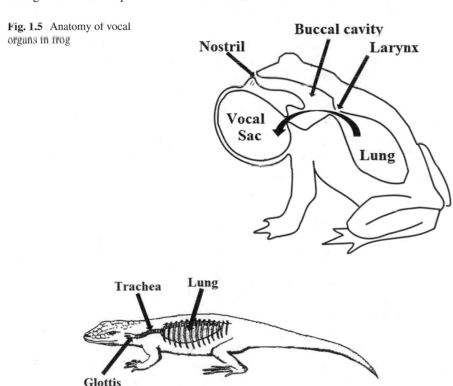

Fig. 1.5 Anatomy of vocal organs in frog

Fig. 1.6 Anatomy of vocal organs in lizard

sound production. Males of Nile crocodiles are best known for 'roar' in order to rival their enemies during territorial disputes. Female crocodiles also can produce similar calls [8].

1.4   Primary Factors in Vertebrate Vocalisation-Based Applications

Data collection is the most challenging aspect when considering acoustic-based applications of vertebrates. Unlike humans and a few domestic animals, researchers cannot hold control of vertebrate vocalisation [9, 10]. Thus, it is a natural process that requires plenty of patience during the vocalisation recordings. In early days, ecologists spent hours and hours in the field listening to sounds of species and collecting audio data for their work, which is a most tedious process. Then keeping microphones in front of cages and making use of laboratory conditions are the techniques used. However, it may not be useful to studies such as behaviour analysis, since it affects the nature of the animal.

Later, the audio sound recording process got new aspirations due to the evolution of modern technologies that are embedded with recording hardware devices. There is a vast range of microphones that are used to collect the audio data, that is, multi-directional, circular and array-type portable microphones [11], Moreover, few studies used sensors that are attached with microphones to collect data [12].

Recent studies have widely approached the autonomous recording units [13, 14]. Although the deployment of such recording units has become more popular and reliable. Nevertheless, several essential factors shall be considered during these deployments. First and the foremost factor is that the unit should not interrupt the natural behaviour of the vertebrates. The second factor is the units should be able to withstand attacks from animals, natural hazards, climatic changes and power supply issues. The general behaviour of vertebrates is not always stable. They tend to move from one place to another since mobility has a significant influence on how an animal obtains nutrients for growth and reproduction. Therefore, it is essential to consider their non-stable nature whilst deploying recording devices.

Furthermore, it is crucial to consider their capability to estimate population area with sound localisation [12]. The up-to-date trend is moving towards fitting equipment (i.e. collars) to animals that can record the sounds and transmit to the base station for further processing [15]. However, many acoustic recordings are available in several reliable online libraries and repositories. Anyone can access these sound recording data free or based on contributors requests [16, 17].

The data labelling is majorly concerned in species-dependent applications and supervised data analysing approaches. Although there are several unsupervised data analysing applications (without labelling) available, yet the ground truth shall be considered during the data processing. The conventional labelling process is carried by the domain experts. They spend long hours listening to the recorded sounds to identify each sound. The sound labels are manually determined to improve the quality of the distant sound analysing applications. However, several forums and social networks are available, where domain knowledge and related vocalisation data can be shared among anyone [17]. Thus, due to globalisation and online networking, domain expert consultation is not a big deal in a technologically driven world. This allows anyone who is interested in developing acoustic-based applications to gather the required sound recordings and apply them to proper preprocessing algorithms to reduce the high-frequency noises during sound recording. They can further perform the other necessary feature analysis and other implementation steps to achieve their specific study aims. However, it is recommended to involve the expertise in these studies to further elaborations such as vocal sound identification and species verifications.

As described in Sect. 1.1, the vocalisation of vertebrates happens in a natural environment, which may include many unavoidable sounds during the recordings, that is, geophony, anthrophony and other biophony sounds [1, 18]. All these types of sounds are considered as noisy sounds that may disturb the essential information of the vocalisation sounds of the vertebrates. Thus, prior to addressing the required acoustic-based applications, sound filtering must be considered.

The critical challenge of sound filtering is to specify the vocals of the targeted sounds since there is a high probability of recording multiple species sounds simultaneously along with the main focused sound. Moreover, the anthropogenic (human-made) noise has made a massive impact in the acoustic environment both on land and water. Many studies determined the consequences of the noise pollutants towards the animals, including vertebrates [19, 20].

The critical information and sound properties can extract easily from the filtered sounds. This information and properties are commonly known as acoustic features. The extraction of these acoustic features plays a significant role to obtain accurate outcomes from its applications. The usefulness of acoustic features and feature extraction methodologies is further detailed in Chap. 2. The modern technological aspects that can be incorporated to invent many novel vertebrate's vocalisation-related applications are further discussed in Chap. 3.

1.5   Acoustic-Based Applications of Vertebrate

Vertebrates create vocal sounds due to a variety of reasons (i.e. social communication, the attraction of opposite gender). The vocal sound analysis of vertebrates may provide enormous information about their breed, sound quality, vocal disorders, age, gender and intercourse. Finding this information can solve several analytical and scientific problems of the animal scientist, biologists and medical persons. Vocal, acoustic feature analysis has been a popular research area from many years back. Especially acoustic analysis on vertebrates is a prevalent and vast study area that can cover various applications such as species identification, communication, behavioural analysis, emotion recognition, species counting and many more.

Nowadays, many organisations have stated that species conservation is essential to prevent many threats to the ecosystem. Especially illegal animal trade, illegal wild animal hunting and environmental pollution cause severe damages to the ecosystem and its biodiversity. It is important to conserve the precious animal species by humankind since humans are responsible for destroying over 80% of wild animals. Improvement of the communication between human to another human and communication of other species among their community and external communities, especially from human to other species communication and vice versa, can create a considerable impact to avoid many social, economic and environmental problems to build a strong relationship among themselves. A recent study on investigating the significance of bioacoustics attempted to solve many challenging tasks that are related to human and animal, including amphibians, birds, fish and reptile communication [21]. Moreover, many economic applications such as ecotourism, manufacturing, cosmetics and food production use a large number of species. Thus, this needs to invest a tremendous amount of attention in educating the people about the behaviour of these species.

Vocalisation is the best media that can give information about social groups, physical and psychological state and mood changes. Many studies focus on vertebrate vocalisation because vocals are the best-heard sounds of any species which give a sense of their behaviours and emotions [14, 22, 23, 24, 25]. Some studies examined the acoustics of the vocals of species to identify different breeds in the same community [18, 15, 26, 27], whilst some studies are focused on investigating the difference between one specie to another [14, 28, 29]. Among various acoustic applications, the species location investigation and individual counting take more attractions in recent research studies [16, 30]. The development of technologies and computer-based machine learning algorithms can turn a new chapter in acoustic analysing applications, mainly to vertebrates because they are the most important species in the ecosystem, specifically for the recreation.

Vertebrate vocal acoustic analysing is a broad range of applications that concerned both active and passive acoustic monitoring strategies. These two strategies are mainly applied to develop the existing instruments and technologies, sound recording instruments and sound feature classification technologies. The instruments and technologies that are integrated with active and passive acoustic monitoring can contribute to improving the knowledge of ecological and geological processes by using the characteristic of different vertebrate sounds. Nevertheless, this book provides both active and passive acoustic monitoring applications, which are based on real-time and pre-recorded sounds, respectively. The overview of this section is drawn on the search of the Google Scholar science journals within the subject categories of 'amphibians', 'fish', 'reptiles', 'birds' and 'mammalians' from 2010 to 2020. The 'call', 'song' and 'vocal'-based sounds are considered to identify their potential applications. Over 40 research articles that are related to the above subject categories are reviewed in this section. The acoustic-based applications of each vertebrate category are elaborated further in below subsections.

1.5.1 *Amphibians*

Amphibians are considered as the most vulnerable species which become endangered in recent years due to global climatic changes [31]. Thus, the applications related to the monitoring of bioacoustic amphibians is vital, yet it is challenging due to the environmental and climatic parameters being influenced on species. Amphibians are bioindicators, who help the scientists to measure the impacts of using chemicals in farming and pollution of river banks [18].

The most common amphibian that can be found in both dry land and freshwater is the frog. The frog produces calls in different scenarios, that is, during their courtship, to notify their location to the others. It shows that the acoustic detection breeding phenology can be identified by the male frog's calling behaviours [31]. Investigation of the breeding phenology may lead to effective conservation strategies and effectively help in population monitoring. Willacy et al. [31] further investigated the breeding phenology of frogs. They introduced an automated recording system along with a frog call recognition software, which helps to monitor the calling occurrence in different environmental temperatures. Dayou et al. [26] classified and identified nine frog species from one family called Microhylidae in real time using their sounds.

The vocalisation of the extant order of the amphibians (anuran) tends to be quiet. This is a disadvantage for many acoustic feature analysing applications on amphibians. Due to this reason, there is a considerable demand to improve an active acoustic feature analysing method to increase the quality of the up-to-date research works. Alonso et al. [32] introduced a sound enhancement method along with an automated classification system for 17 anuran species, whilst Noda et al. [33] differentiate and identify anuran species from reptiles using various classification methods. Tomasini et al. [18] implemented a smartphone application which can classify and identify calls of 48 anuran species, which are collected from 50 locations all over the United States. Furthermore, they extended their findings to monitor the anuran population.

Moreover, Alberto da Silva et al. [34] proposed an energy-efficient hardware implementation called Mana-M. This hardware implementation includes an acoustic sensor node to monitor and classify anurans, which is the initial prototype to identify only one type of anuran species. This achievement was considered as a proof of concept on their research work. Thus, they stated that their work should need more improvements, including additional functionalities to identify various anurans.

1.5.2 *Fishes*

The acoustics of fishes differ between the types of species and their population, size and gender. Over 800 species belonging to more than 100 families are documented as sound-producing fishes. The acoustic characteristics of fish were influenced by motivation as well as fluctuations in the surrounding environment [19]. The acoustic features of fishes can address various applications such as communication between species, detection ability of prey and predators, habitat selection, mating behaviour, orientation and migration pattern. There are two types of behavioural contexts, that is, aggression and reproduction. These behavioural contexts are often associated with the sounds that fish produce. Typically, during the breeding seasons, fishes will produce more intense sounds. However, humans barely heard these sounds because the majority of these sounds are having lower frequency ranges, usually less than 1000 Hz. Thus, it shall note that divers can hear some sounds that are composed by reef fish category at deep inside the water [24].

In the field of fisheries science, determining when and where fishes spawn is a crucial aspect. Earlier 'spawn-o-meter' is used to find spawning sites [35]. Spawning information provides essential data on habitats used by the species and will lead to better management of threatened species for conservation purposes. However, an increase of the ambient noise, mostly anthropogenic noise, interferes fishes' communication and affects its ability to detect and respond to relevant sounds. It will affect the survival of individuals and the population [7].

The impact of acoustic-based monitoring is a difficult and challenging task [19]. Lindseth et al. [35] reviewed a few research studies that applied advanced audio signal processing technologies to contribute to the underwater soundscape monitoring by detecting and counting the fishes since fish produces a unique sound when spawning. Rowell et al. [36] proposed an acoustic level-based model to estimate the abundance, density and biomass of fishes at the spawning aggregation by analysing the unique sounds produced by fishes during spawning. A multi-method approach for passive acoustic fish monitoring is introduced by Lin et al. [37] to monitor the chorusing events of soniferous fish.

1.5.3 Reptiles

In the visibility, limited remote environment bioacoustic-based species identification is an effective way to sample population and control conservation of small living beings such as reptiles. The vocalisation of the reptiles, including vocal plasticity, is more flexible than many other vertebrates. They can produce sound syllables with higher amplitudes and have the ability to increase the duration of their calls [38].

Among many varieties of the reptile family, sea turtle hatchlings are well known for vocalisations. Lindsay et al. [39] observed the vocalisation frequency, which may help to identify the habitats of sea turtles. This vocalisation frequency was further examined to understand the behaviour of their hatching vocalisations, which includes synchronising hatching and starting the emergence from the nest.

Crocodilians, dinosaurs and pterosaurs are a few of the common reptiles that can be considered for parental care. According to Chabert et al. [40], parental care was defined differently. They have mentioned the acoustic interaction is vital when considering the relationship between offspring and mother. Hence, the changes in communication may affect maternal care, and it may lead to dispersal of offspring. Chabert et al. [40] further examined the response of a pregnant Nile crocodile to offspring during its breeding season. They found that the Nile crocodile mother is less receptive to larger juvenile calls. They further justified their findings by stating that a variation in the acoustics of juvenile call is based on its size that can be recognised by the mother crocodile. Addition to these findings, Mathevon et al. [41] examined the acoustic communications of juvenile crocodiles to understand the interspecific differences of their calls, whilst Noda et al. [33] differentiated the reptiles and anuran species using a fusion of significant properties that are present in reptile vocalisation.

1.5.4   Birds

Bird sound-related research focuses mainly on species identification and estimation of bird species richness in particular areas, assessing the status of threatened species and monitoring of climatic changes through bird migration pattern and early warnings for the natural hazards [16]. There are over 18,000 identified bird species in the world, and many of them are region-based species that can be found in specific regions or continents. Thus, South East Asia has over 1700 bird species and ranked as the fourth place among all the other regions. In their research, Ramashini et al. [16] considered the calls of birds in the Borneo region, which is considered as a part of South East Asia. They suggested a simple linear classification-based bird identification method with high classification accuracy. This study is stated as the first-ever audio-based bird identification study for endemic forest Bornean birds. The Gaga' chicken crowing sound is compared with other singing chicken in Indonesia and Japan by Bugiwati et al. [42]. This study identified that the Gaga' chicken has a crowing sound which is very specific, unique and different than the other chickens. The ending of the crowing sound is very similar to the human laughing voice characteristics.

A vast range of applications on bird classification has been carried out over the years. Among many other bird acoustic-based applications, the birds' sound recording and counting get more attention. Stattner et al. [12] invented a hardware-based approach to count the singing birds automatically. In the bird's habitat, a wireless sensor fitted with a microphone is kept to collect audio data samples. Further, the song fingerprint is extracted and identified individual species by a classification method. Where else, Padovese et al. [17] suggested a machine learning approach for bird survey using an automated acoustic landscape monitoring systems. This study tried to recognise the Brazilian Atlantic Forest parrot species.

1.5.5 Mammalians

Mammalians are the species that have a backbone and hair in the body. There are more than 6000 species all over the world. They can be categorised into three main groups such as monotremes, marsupials and placental mammals [43]. The monotreme mammalian group is known to lay eggs. Currently, there are only a few types of monotremes that live in the world (i.e. spiny anteater, or echidna, and platypus). They can be commonly found in the Australian continent. Compared to the other types of mammals, the monotremes have different body features with a lower body temperature level [43]. Marsupial mammals give birth to babies, yet the babies are not fully developed. Then the babies will grow up inside a pouch on the outside of their mother's abdomen by drinking milk from the mother. The most well-known marsupial mammals are kangaroos, koalas, opossums and wallabies. They are commonly found in both Australian and South American continents [43]. Placental mammals are the most common mammals, including humans, that can be found in both water and land. Unlike the marsupial mammals, this type develops inside the mother's body until the body structure and the functions are stable [43]. However, similar to marsupial mammals, the placental mammals also have their mother's milk to intake nutrition to improve their body functionalities.

The applications which are based on the acoustics of the mammals are a highly complicated and vast study area. There are many acoustic-based applications such as acoustic processing, speech processing, biomedical signal processing, sound recognition, sound localisation, robotics and AI that are highly focused as the current research scope. Apart from these applications, the voice-related applications can be named as the top-ranked acoustic feature analysing applications.

The voice-related bioacoustic applications on mammals are mostly concerned to identify their behaviours. Notably, that can be applied to understand the feelings of the domestic mammals, that is, cats and dogs. Nicastro et al. [44] analysed and compared the acoustic characteristics of the 'meow' sound of both domestic and African wild cats (cat crying sound) to identify the similarities between them. They found that the domestic cat 'meows' are considerably shorter than the wild cats. Furthermore, they assure that the 'meow' sound of the domestic cats is more pleasant than the wild cats. Pandeya et al. [45] constructed a generalised robust domestic cat sound database using audio data augmentation to grip variations of different cat species, the variation of their age and biodiversity. Riede et al. [46] recorded the 'growling' sound of domestic dogs to analyse the length of recorded vocal tracks. Moreover, they used the formant dispersion of the domestic dogs to extract the information of the body size of the vocaliser. Taylor et al. [47] suggested that analysing the vocaliser size of the dogs using the formant dispersion may provide the behavioural response of the body size of each domestic dog separately, which can elaborate further as the small-sized dogs' vocal response is less than the huge-sized dogs. Frommolt et al. [10] investigated the directionality patterns of the vocal sounds of the domestic dogs to understand their sound structure and behaviour during the vocalisation.

Similar to the acoustic-based applications of domestic mammals, the wild mammals have been considered in some past research studies. Garcia et al. [48] identified the vocal learning capabilities of 164 mammals by studying acoustic outliers. The bioacoustic analysis is mainly studied for three types of wild mammals: Elephantidae, Chiroptera, and Cetacea. Elephantidae are named as the largest herbivorous mammals in the land, which include elephants and mammoths. Research studies are conducted to understand the vocalisation of the elephants because they have been identified as reserved mammals in the land due to the decrease in their populations. Venter et al. [49] suggested an algorithm to detect elephant vocalisations from noisy signals. Clemens et al. [15] classified the vocalisation of five different African elephants. Angela et al. [27] analysed the vocal imitation and sound production mechanism to investigate the similarities and differences of the communication between African and Asian elephants.

Chiroptera is a particular type of mammalian who has a body structure which is different from other mammals. They are capable of flying as birds due to their forelimbs which are adapted as wings. Bats are the well-known Chiroptera mammals that have noticed in the bioacoustic field of study. Bats are one of the rare types of mammals which have their way of navigation using echolocation. The echolocation is a vivid technique to estimate the distance for an obstacle by analysing the reflection of the sound. Prior to flight, they send a sound and wait for its reflection.

Bats are known to have advanced ears among all the mammals. Thus, they adapted echolocation to create a form of vision in the dark to avoid the obstacles in flight and find food. Britzke et al. [30] recorded echolocation calls of different types of bat species in the Eastern United States to classify them into different categories by analysing the characteristics of calls. Jones et al. [50] analysed the echolocation pulse of different bat species to identify each species category and to differentiate male and female. Prat et al. [51] analysed behaviour response of the calls and call's addressee of Egyptian fruit bats. Moreover, in their study, they investigated the everyday interactions of the bats by analysing the same call vocalisations.

Cetacea mammalians is another name for aquatic or marine mammals. There are two types of Cetacea: Odontoceti and Mysticeti. The Odontoceti mammals include the dolphin and few types of whales, that is, beaked whales and sperm whales, whilst Mysticeti mammals include grey whale, the blue whale and many more, which contain a filter feeder system. Compared to the marine species, whales and dolphins are considered as the most intelligent species. People use these mammals in sea navigation and whale watching to earn more economic benefits. Jones et al. [25] provided a detailed literature review on dolphin sounds to build tools to recognise unique sound styles and understand their communications and behaviours. Fauda et al. [13] investigated the influence of concurrent ambient noise level, which is caused by the ship to the whistle calls of bottlenose dolphins in the Western North Atlantic Ocean. They have further elaborated the proposed application to identify the impact of ambient noise to dolphin-to-dolphin communication. Musser et al. [28] proposed an acoustic feature-based classification method to differentiate bottlenose dolphins and killer whales. Antunes et al. [14] analysed the vocals of sperm whales to identify their relationship and behaviours towards the other members of the same unit. In addition to these findings, they have investigated the individual behaviours of the sperm whales by testing the variation within coda types.

Humans are stated as the most important and unique mammals in many ways. Humans also share many features as same as the other mammals such as growing hair from the skin, production of milk to feed the babies and many more. Compared to all the known animals, humans have the highest vocal production learning (VPL), which can develop their communication skills through various methods and advanced technologies. Thus, they can express and understand any feelings and can respond quickly. The acoustic feature analysis is famous in many applications, mostly in medical and biological applications. Apart from these two applications, human vocalisation analysis is applied in military, crime investigations, telecommunication and economic applications. Almost all the vocalisation applications are based on speech signal processing, which may be considered as a part of voice analysis. Hansen et al. [23] studied the acoustic features and changes of the vocal effort of loud scream, which is produced due to anger, distress and fear, to verify the identity of the speaker. The recent evidence shows that the most important human auditory applications are based on speech analysis because many technologies and applications have developed to operate as per the voice commands of the users. Mekyska et al. [9] used human speech recordings to improve medical diagnosis. In their work, they identified many speech pathologies using parameterisation techniques, which have adapted by analysing many voice-related features. Additionally, they have introduced 36 novel pathological voice measures based on various speech-related features. Mao et al. [22] developed a robust computer-based application to understand the human emotional behaviours by investigating their speech signals. This application may assist medical applications to diagnose many psychological and personality disorders. Moreover, speech emotional recognition (SER) will give a high impact in education and customer service and understand the human emotional state clearly.

Beside the Elephantidae, Chiroptera, Cetacea, and humans, the bioacoustic features are applied to many other mammals who are commonly heard in the society. Petersen et al. [29] compared the acoustic communications of European wolves and various dog breeds to understand the behavioural and function response as well as their emotions of all significant sounds. In their dissertation, Hooper et al. [20] discussed the impact of the development of bioacoustic tools and applications for mammals. They have further described the use of bioacoustic tools that can apply to the alarm calls of Belding's ground squirrel to get the information that is related to their age and gender. Moreover, they investigated the impact of road noise on Belding's ground squirrel alarm calls. Benjamin et al. [11] identified the synchronising mating behaviour of vocalisations and acoustic variation of the calls of giant pandas, who are a member of the bear family. These pandas are considered to be in the category of Carnivora mammals, who only depend on a vegetarian diet. In addition to these applications, the behavioural analysis of chimpanzees is widely studied because they are considered as the closest relatives to the human being. This relationship may be applied in vocal pattern analysis since chimpanzees share about 99% similar deoxyribonucleic acid (DNA) with humans. Taglialatela et al. [52]

analysed the laughter of chimpanzees to examine the similarities of nonhuman and human laughter in terms of emotional recognition since humans express their emotions through laughing. They have found that chimpanzees have substantial communicative advantages and emotional expressions among their community by responding with laughter. Correspondingly, these findings prove that the chimpanzees have similar vocalisations to human vocalisations.

1.6 Summary

The vocal sound production of vertebrates is considered as the most definite form of communication. Vocal communication which includes 'calls' and 'songs' is mainly used for various purposes such as location identification, food search, obstacle detection, warning calls and courtship. The production of vocal communication is referred to as vocalisation. Many living creatures express their emotions through vocalisation. Thus, the importance of analysing the vocalisation of vertebrates shall be beneficial to a vast range of applications such as tourism, economy, medical, biology, geology and animal science.

Vocalisation analysis of vertebrates contains a massive range of challenges, and many of these challenges are still not addressed. The data recording, processing and best acoustic feature selection can be categorised as the most difficult challenges that are faced in past and recent studies. However, there is a high possibility to resolve these challenges by integrating the available technologies. Thus, this integration can improve the quality of data recording instruments and process the recorded data accurately to find novel acoustic features of each vertebrae category. Nevertheless, almost all of the studies still depend on the regional and species-based acoustic pre-recorded data, which is commonly used in passive acoustic monitoring applications. However, there is a vast spectrum of work that needs to be addressed in active acoustic monitoring-based applications.

This chapter has revealed the vertebrate vocalisation-based applications that must need more improvements in the future. These improvements can be made by analysing the different acoustic features of sounds and categorising their similarities and differences between each vertebrate class. Incorporation of modern technologies can be further integrated to develop fascinating vertebrate acoustic-based inventions to protect biodiversity and ecosystem.

References

1. B.C. Pijanowski, A. Farina, S.H. Gage, S.L. Dumyahn, B.L. Krause, What is soundscape ecology? An introduction and overview of an emerging new science. Landsc. Ecol. **26**(9), 1213–1232 (2011)
2. G. Manteuffel, B. Puppe, P.C. Schön, Vocalisation of farm animals as a measure of welfare. Appl. Anim. Behav. Sci. **88**(1–2), 163–182 (2004)

3. L.R. Hernandez-Miranda, C. Birchmeier, Mechanisms and neuronal control of vocalisation in vertebrates. Opera Med. Physiol. **4**(2), 50–62 (2018)
4. C. Gans, C.J. Bell, Vertebrates, Overview, in *Encyclopedia of Biodiversity*, 2nd edn., (Elsevier Inc, 2001), pp. 333–341
5. R.A. Suthers, *How Birds Sing and Why It Matters* (2004)
6. M.J. Ryan, M.A. Guerra, The mechanism of sound production in túngara frogs and its role in sexual selection and speciation. Curr. Opin Neurobiol. **28**. Elsevier Ltd, 54–59 (2014)
7. A.N. Popper, A.D. Hawkins, An overview of fish bioacoustics and the impacts of anthropogenic sounds on fishes. J. Fish Biol. **94**(5), 692–713 (2019)
8. C. Gans, P.F.A. Maderson, *Sound Producing Mechanisms in Recent Reptiles: Review and Comment* (1973)
9. J. Mekyska et al., Robust and complex approach of pathological speech signal analysis. Neurocomputing **167**, 94–111 (2015)
10. K.-H. Frommolt, A. Gebler, Directionality of dog vocalisations. J. Acoust. Soc. Am. **116**(1), 561–565 (2004)
11. B.D. Charlton, M.S. Martin-Wintle, M.A. Owen, H. Zhang, R.R. Swaisgood, Vocal behaviour predicts mating success in giant pandas. R. Soc. Open Sci. **5**(10) (2018)
12. E. Stattner, N. Vidot, P. Hunel, M. Collard, Wireless sensor network for habitat monitoring: A counting heuristic. Proc. - Conf. Local Comput. Networks, LCN, 753–760 (2012)
13. L. Fouda et al., Dolphins simplify their vocal calls in response to increased ambient noise. Biol. Lett. **14**(10) (2018)
14. R. Antunes, T. Schulz, S. Gero, H. Whitehead, J. Gordon, L. Rendell, Individually distinctive acoustic features in sperm whale codas. Anim. Behav. **81**(4), 723–730 (2011)
15. P.J. Clemins, M.T. Johnson, K.M. Leong, A. Savage, Automatic classification and speaker identification of African elephant (Loxodonta africana) vocalisations. J. Acoust. Soc. Am. **117**(2), 956–963 (2005)
16. M. Ramashini, P.E. Abas, U. Grafe, L.C. De Silva, Bird Sounds Classification Using Linear Discriminant Analysis. ICRAIE 2019 - 4th Int. Conf. Work. Recent Adv. Innov. Eng. Thriving Technol. **2019**(November), 27–29 (2019)
17. B.T. Padovese, L.R. Padovese, *A Machine Learning Approach to the Recognition of Brazilian Atlantic Forest Parrot Species* (2019)
18. M. Tomasini, K. Smart, R. Menezes, M. Bush, E. Ribeiro, Automated robust Anuran classification by extracting elliptical feature pairs from audio spectrograms, in *ICASSP, IEEE International Conference on Acoustics, Speech and Signal Processing - Proceedings*, vol. 1152306, (2017), pp. 2517–2521
19. A.N. Radford, E. Kerridge, S.D. Simpson, Acoustic communication in a noisy world: Can fish compete with anthropogenic noise? Behav. Ecol. **25**(5), 1022–1030 (2014)
20. S.L. Hooper, *Impacts and Applications Developing a Bioacoustic Tool for Mammals and Measuring the Effects of Highway Noise on a Mammalian Communication System, Using Ground Squirrels as a Model* (2011), p. 102
21. P. Laiolo, The emerging significance of bioacoustics in animal species conservation. Biol. Conserv. **143**(7), 1635–1645 (2010)
22. Q. Mao, M. Dong, Z. Huang, Y. Zhan, Learning salient features for speech emotion recognition using convolutional neural networks. IEEE Trans. Multimed. **16**(8), 2203–2213 (2014)
23. J.H.L. Hansen, M.K. Nandwana, N. Shokouhi, Analysis of human scream and its impact on text-independent speaker verification. J. Acoust. Soc. Am. **141**(4), 2957–2967 (2017)
24. P.S. Lobel, I.M. Kaatz, A.N. Rice, Acoustical behavior of coral reef fishes. no. March 2016 (2010)
25. B. Jones, M. Zapetis, M.M. Samuelson, S. Ridgway, Sounds produced by bottlenose dolphins (Tursiops): A review of the defining characteristics and acoustic criteria of the dolphin vocal repertoire. Bioacoustics **29**(4), 399–440 (2020)
26. J. Dayou, N.C. Han, H.C. Mun, A.H. Ahmad, Classification and identification of frog sound based on entropy approach. Ipcbee **3**, 184–187 (2011)

27. G. Witzany, *Biocommunication of Animals* (2013)
28. W.B. Musser, A.E. Bowles, D.M. Grebner, J.L. Crance, Differences in acoustic features of vocalisations produced by killer whales cross-socialised with bottlenose dolphins. J. Acoust. Soc. Am. **136**(4), 1990–2002 (2014)
29. D.U. Feddersen-Petersen, Vocalisation of European wolves (Canis lupus lupus L.) and various dog breeds (Canis lupus f. fam.). Arch. Anim. Breed. **43**(4), 387–397 (2000)
30. E.R. Britzke, K.L. Murray, J.S. Heywood, L.W. Robbins, Acoustic identification. Indiana Bat Biol. Manag. Endanger. Species, December 2015, 221–225 (2002)
31. R.J. Willacy, M. Mahony, D.A. Newell, If a frog calls in the forest: Bioacoustic monitoring reveals the breeding phenology of the endangered Richmond range mountain frog (Philoria richmondensis). Austral Ecol. **40**(6), 625–633 (2015)
32. J.B. Alonso et al., Automatic anuran identification using noise removal and audio activity detection. Expert Syst. Appl. **72**, 83–92 (2017)
33. J.J. Noda, D. Sánchez-Rodríguez, C.M. Travieso-González, *A Methodology Based on Bioacoustic Information for Automatic Identification of Reptiles and Anurans* (Reptil. Amphib, 2018)
34. C. Alberto da Silva, L. Beatrys Ruiz, MannAnuro: Classification and identification of anuran amphibians using wireless multimedia sensor network. IOSR J. Comput. Eng. Ver. I **17**(5), 2278–2661 (2015)
35. A.V. Lindseth, P.S. Lobel, Underwater soundscape monitoring and fish bioacoustics: A review. Fishes **3**(3) (2018)
36. T.J. Rowell, D.A. Demer, O. Aburto-Oropeza, J.J. Cota-Nieto, J.R. Hyde, B.E. Erisman, Estimating fish abundance at spawning aggregations from courtship sound levels. Sci. Rep. **7**(1) (2017)
37. T.-H. Lin, Y. Tsao, T. Akamatsu, Comparison of passive acoustic soniferous fish monitoring with supervised and unsupervised approaches. J. Acoust. Soc. Am. **143**(4), EL278–EL284 (2018)
38. H. Drumm, S.A. Zollinger, Vocal plasticity in a reptile. Proc. R. Soc. B Biol. Sci. **284**, 2017 (1855)
39. N.J.R. Lindsay, N. McKenna, F.V. Paladino, P.S. Tomillo, Do sea turtles vocalise to synchronise hatching or nest emergence? Am. Soc. Ichthyol. Herpetol. **107**(1), 120–123
40. T. Chabert et al., Size does matter: Crocodile mothers react more to the voice of smaller offspring. Sci. Rep. **5**, 1–13 (2015)
41. N. Mathevon, A. Vergne, T. Aubin, Acoustic communication in crocodiles: How do juvenile calls code information? Proc. Meet. Acoust. **19**, 1–5 (2013)
42. F. Bugiwati, S.R. Aprilita, Ashari, Crowing Sound Analysis of Gaga ' Chicken: Local Chicken From South Sulawesi Indonesia Faculty of Animal Husbandry , Hasanuddin University, Jl. Perintis Kemerdekaan Km. 10, Tamalanrea, Makassar (90245), South Sulawesi, Indonesia Phone : 0411-58311, no. April, pp. 163–168, (2013)
43. Naturalsciences.ch, "Three Types of," no. 1989, p. 1 (2017)
44. N. Nicastro, Perceptual and acoustic evidence for species-level differences in meow vocalisations by domestic cats (Felis catus) and African wild cats (Felis silvestris lybica). J. Comp. Psychol. **118**(3), 287–296 (2004)
45. Y.R. Pandeya, J. Lee, Domestic cat sound classification using transfer learning. Int. J. Fuzzy Log. Intell. Syst. **18**(2), 154–160 (2018)
46. T. Riede, T. Fitch, Vocal tract length and acoustics of vocalisation in the domestic dog (Canis familiaris). J. Exp. Biol. **202**(20), 2859–2867 (1999)
47. A.M. Taylor, D. Reby, K. McComb, Size communication in domestic dog, Canis familiaris, growls. Anim. Behav. **79**(1), 205–210 (2010)
48. M. Garcia, A. Ravignani, Acoustic allometry and vocal learning in mammals. Biol. Lett. **16**(7), 20200081 (2020)
49. P.J. Venter, J.J. Hanekom, Automatic detection of African elephant (Loxodonta africana) infrasonic vocalisations from recordings. Biosyst. Eng. **106**(3), 286–294 (2010)

50. G. Jones, B.M. Siemers, The communicative potential of bat echolocation pulses. J. Comp. Physiol. A Neuroethol. Sensory, Neural, Behav. Physiol. **197**(5), 447–457 (2011)
51. Y. Prat, M. Taub, Y. Yovel, Everyday bat vocalisations contain information about emitter, addressee, context, and behavior. Sci. Rep., November **6**, 1–10 (2016)
52. J.P. Taglialatela, J.L. Russell, J.A. Schaeffer, W.D. Hopkins, Chimpanzee vocal signaling points to a multimodal origin of human language. PLoS One **6**(4), 1–7 (2011)

Chapter 2
Significance of Acoustic Features in Vertebrate Vocalisations Related Applications

2.1 Introduction to Acoustic Features

The acoustic features are considered as a property of a sound wave. These types of features are usually applied in auditory signal processing that includes speech signals, music, voice and many other audio related signals. The characteristic analysis of acoustic features is applied in many fields, ranging from linguistics to machine recognition [1]. However, the vertebrate vocalisation involves two separate processes such as initial sound and modified sound. When the vertebrate produces an initial sound which shall contain various frequencies in the sound, thus in the modifying process, they are eligible to use tongue, lips or teeth to modify the spectrum of the initial sound over time [2].

Vertebrates including mammals, amphibians, reptiles, fish and birds produce different sounds through their vocal codes, but many of these sounds such as calls and speech are produced for the communication and emotion recognition tasks. Acoustic features carry meaningful information; thus, the feature extraction and selection is an extremely challenging task [3], especially in vertebrate vocalisation analysis. The importance of extracting the suitable acoustic features is essential in most animal welfare and health-related research studies as well as to improve the various types of environmental factors such as stocking density, habitat and disease prevention [3]. However, the rapid increase of advanced technologies takes a major role in analysing the acoustic features as an aid to animal conservation and welfare applications. Therefore, it is vital to extract and select the best acoustic features of the vertebrate vocalisations using the most technically advanced algorithms to achieve the best outcome from the application.

Acoustic features can designate for different categories, namely, time related, frequency related, cepstral and temporal. Energy and frequency of the audio signals can create different acoustic patterns [4]. There is a significant acoustic transition from one sound to another sound in vertebrate vocalisation, which can vary the

R. Murugaiya et al., *Acoustic-Based Applications for Vertebrate Vocalization*, SpringerBriefs in Applied Sciences and Technology, https://doi.org/10.1007/978-3-030-85773-8_2

location and shape of the vocal patterns. The information of these vocal patterns shall be stored as the acoustic features to make the primary voice identification applications that include vertebrate's call identifications, human speech recognition, voice pathology recognition as well as vertebrate emotion recognition. For this purpose, the uniqueness of the acoustic characteristics of vertebrate vocal sounds may be used to establish a perceptual unit in the primary voice recognition process.

2.2 Related Works on Acoustic Features

Most of the studies have investigated the effect of the vertebrate vocalisations through the number of acoustic features and their articulatory and perceptual parameters. Hansen et al. [5] considered the frequency-related features, that is, Mel frequency cepstral coefficient (MFCC), perceptual minimum variance distortionless response (PMVDR) and energy distribution of the human neutral speech and scream signals to observe their pattern differences. They extracted the MFCC features via Mel scale filter bank to the magnitude spectrum of short-term fast Fourier transform (STFFT) and a spectrum which is based on the short-term linear predictive coding (STLPC) to achieve a perceptually relevant smoothed gross spectrum. Similarly, they have extracted the cepstral coefficients of the human neutral speech and scream signals via PMVDR by integrating a perceptual deformation of the FFT power spectrum. Mcloughlin et al. [3] focused on the acoustic features that are related to both time and frequency domains such as MFCC, LPC, Mel spectrogram, fundamental frequency, spectral centroid, spectral flux, spectral flatness and zero-crossing rate to classify the animal vocalisation in order to identify the different species and estimate the numbers of individuals.

Furthermore, Clemins et al. [6] applied a moving hamming window to extract the MFCC and log energy features of the African elephants' call sounds, whilst Venter et al. [7] applied only MFCC features to classify the vocalisations of African elephants using hidden Markov models. Also, Owren et al. [8] extracted the acoustic features such as LPC, cursor-based measurements of centre frequency, overall slope of the frequency spectrum, means and standard deviations of fundamental frequency values (taken from time waveform), first partial value, the interval between partial and amplitude of vowels in grunt calls of chacma baboons. They have extracted these acoustic features for 216 chacma baboon's call recordings to investigate the individual call patterns and identify the female baboon's vocal sounds.

Similar to land mammals, marine mammals often produce stereotypic sounds, which have many variations between individuals [9]. Melissa et al. [10] applied both time-dependent and frequency-dependent acoustic features to identify the aggressive chase in the killer whale. They used the features such as overall call length, the number of segments, overall amplitude variation, rhythmic amplitude modulation, average slope, the preponderance of rhythmic frequency modulation (RFM) and variation in RFM as time-dependent features, while the number of harmonics and sidebands, average distance between second and third sidebands, degree of RFM, slope differences among sidebands, RFM differences among sidebands and frequency range within sidebands as the frequency-dependent features. Subsequently, they have concluded the selected features can provide better results in identifying the frequencies and durations of the calls of killer whales.

The combination of different texture features with acoustic features not only is equivalent to current vertebrate audio signal approaches but also statistically enhances some of the stand-alone audio features. Therefore, Nanni et al. [11] plotted the vocal signals of birds and whales as a spectrogram image to extract the visual representation features, that is, amplitudes, statistical spectrum descriptors (SSD), rhythm histogram (RH), modulation frequency variance descriptor (MVD), temporal rhythm histograms and temporal SSD from spectrogram image. Burkett et al. [12] extracted the acoustic features such as mean, pitch and Weiner entropy to cluster the bird song syllables. Ramashini et al. [13] introduced a bird sound classification algorithm, specifically to the birds that can be found in the Borneo rainforest. They have investigated both time and frequency domain features such as zero-crossing rate (ZCR), energy (E), entropy of energy, spectral centroid, spectral spread, spectral entropy, spectral flux, spectral roll-off, MFCCs and chroma vectors. Apart from these features, they have extracted a few other acoustic features including harmonic ratio and fundamental period. Additionally, they have further elaborated the extracted features via linear discriminant analysis (LDA) feature processing tools to reduce the dimensionality of all extracted features and find the best acoustic features by removing the redundant features.

The cepstral features such as MFCC and LPCC are widely used acoustic features in many vertebrate vocalization applications, that is, speech recognition [14–16], call recognition [10, 17, 18] and medical applications [19–21]. Also, these features perform better than the other common acoustic features, that is, time-related features and statistical features. However, the feature extraction task shall provide a strong representation of the vocalisation due to its response on robust vocal characteristics and uniqueness. Therefore, it is necessary to select the finest features and feature extraction methods to emphasise the best results for the applications which are related to the vertebrate vocalisation.

Moreover, many acoustic features are widely used in vocal signal analysis. However, homomorphic signal processing and its derivatives such as cepstral features are the commonly used general techniques in the audio signal analysis that involves nonlinear mapping in different domains. Thus, the cepstral features are the ideal acoustic feature for the applications that are based on the vertebrate vocalisation. The following section, the significance of homomorphic signal processing method in vertebrate vocalisation and its cepstral derivatives are discussed in depth.

2.3   Homomorphic Signal Processing

Vertebrates play a major role in the sustainability of ecosystems. Therefore, vertebrate conservation and species preservation-related projects are mandatory for a peaceful ecosystem. As discussed in Chap. 1, there is a vast range of solutions that can be addressed using vertebrate vocalisation-based applications. However, implementation of these types of applications is considered to be difficult because of the work, labour and physically intensive process. Along with the emergence of modern techniques, the environmental and biodiversity monitoring related applications got a new lease of life. Thus, signal processing and machine learning techniques are used to ease this complicated process, especially in dense vegetation, while bioacoustics signal processing and pattern recognition algorithms are used to detect and identify varieties of species [22].

The vocalisation production process must explore thoroughly to develop vertebrate vocalisation related applications. As discussed in Chap. 1, the vocal organ and respiratory pattern play a vital role in animal vocal production. However, the vocal organs of vertebrates are different from one to another, and it has a high influence in producing different sounds for each vertebrate, although the excitation is the same. Therefore, the excitation is convoluted with the structure of the vocal organs and associated activities to produce the sound [23].

Homomorphic signal processing method and its derived features can be applied in many signal processing applications such as speech recognition, deconvolution, pitch detection and image enhancement. Primarily, the success of homomorphic signal processing is found in echo removal and speech analysis [24]. There are three major steps in homomorphic signal processing methods such as Fourier transform, non-linear transform (usually a logarithm function) and inverse Fourier transform (IFT) [25]. An illustration of these three steps is shown in Fig. 2.1.

In vertebrate vocalisation based applications, the homomorphic signal processing method could be a better option that can be used to deconvolute the signal which is based on the structure of the vocal organ and associated activities h[n], yet this is responsible for the difference in sounds from the excitation signal e[n]. The produced vocal sound s[n] can be represented by:

$$s[n] = h[n] * e[n] \tag{2.1}$$

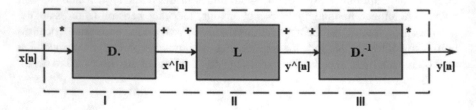

Fig. 2.1 The homomorphic signal processing method

However, in the frequency domain (after the discrete Fourier transform (DFT)), it can be represented as:

$$S[k] = H[k] * E[k] \tag{2.2}$$

where $S[k]$ is the produced vocal sound, $H[k]$ is the signal produced by the structure of the vocal organ and associated activities and $E[k]$ is the excitation signal. Thus, it can take either an absolute log or simple log in log domain by:

$$\log(S[k]) = \log(H[k]) * \log(E[k]) \tag{2.3}$$

$$\log(\|S[k]\|) = \log(\|H[k]\|) * \log(\|E[k]\|) \tag{2.4}$$

As $H(k)$ will become an additive function in the log domain, now it can be extracted, separately [26]. Additionally, the computed additive function can be observed in Fig. 2.2, where the envelope represents as $H[k]$ and variation represents as $E[k]$. However, the centre of attraction is on $H[k]$, which can be passed through the low-pass filter shall be separated [27].

The produced sound is modelled as the output of linear time in varying (LTI) system, excitation and impulse response are combined in a convolution manner. Transforming non-linearly combined signals to addictively combined signals is the major challenge [28]. Once the IFT is taken to the log spectrum, then it can be referred to as cepstrum, and determining the cepstrum is called cepstrum analysis which is illustrated in Fig. 2.3.

As shown in Fig. 2.3, the real cepstrum will be produced only if the log magnitude is considered; thus this may consist of only the magnitude of the complex log that can produce complex cepstrum which consists of phase and magnitude. The

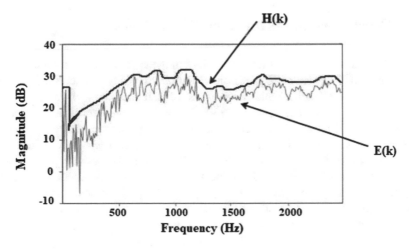

Fig. 2.2 Envelop and variation representation of magnitude spectrum

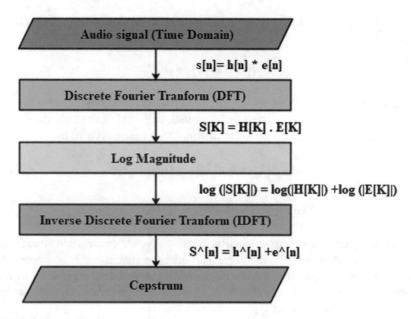

Fig. 2.3 Canonic form for obtaining cepstrum

$h^{\wedge}[n]$ represents the envelope of the spectrum, and this cepstral domain is called the 'quefrency domain'. To do the necessary separations of $h^{\wedge}[n]$, the filtering is a must, and this is known as 'liftering' [29]. The cepstrum can be passed through a low-pass filter to get cepstral coefficients. The beginning is a slowly varying component, so those are called cepstral coefficients. For example, if N length IDFT, then 0 to N/2 coefficients will represent the signal envelope.

2.4 Cepstral Features for Acoustics Analysis

Cepstral features are compact representations of the spectrum and provide a smooth approximation based on the logarithmic magnitude. The cepstral coefficients can be derived in two approaches: one approach is a linear prediction approach, and the other one is the filter bank approach [30]. Linear prediction cepstrum coefficients (LPCC), Mel frequency cepstral coefficients (MFCC), gammatone frequency cepstral coefficients (GTCC), perceptual linear prediction (PLP) cepstral coefficient and Greenwood function cepstral coefficients (GFCC) are few types of cepstral features [31]. It has been primarily used for speaker identification and speech recognition [26]. However, they have also been employed in the applications of audio retrieval such as singer identification, music classification, environmental sound recognition, pitch determination of speech signals and for identification of musical instruments [31, 32].

2.4.1   Linear Prediction Cepstrum Coefficients (LPCC)

The LPCC features will be derived based on the linear prediction approach, which can manifest more useful emotion-specific information through the vocal tract. The basic idea is to estimate the nth audio sample by a linear combination of its previous k samples [30]. Prior to extracting the LPCC, the audio signal pre-emphasis shall be performed. After pre-emphasis, the signal will be framed into small chunks using windowing techniques. Then the auto-correlation will be found to calculate the LPC parameters. Finally, the LPC parameters will be converted into the cepstral domain [31].

2.4.2   Perceptual Linear Prediction (PLP) Cepstral Coefficients

Similar to the LPCC, the PLP feature is also derived from the LPC, where it involves a slightly different process. The signal shall be framed to derive the PLP cepstral coefficients from audio signals by the windowing technique. Then it must perform critical band analysis, equal loudness and pre-emphasis tasks. Intensity-loudness conversion will be done before applying the linear prediction algorithm and finally converted into the cepstral domain [31].

2.4.3   Mel Frequency Cepstral Coefficients (MFCC)

The MFCC is a set of features that are used in many audio-based applications since these features will closely mimic the human auditory system with the frequency bands which are equally spaced on Mel scale. Hence, these features can be extracted with several simple steps, yet as the first step, it still needs to frame the audio signal into short frames using a windowing technique. Then from each frame periodogram estimate of the power spectrum will be calculated and passed through the set of Mel filter banks [30]. The approximation of Mel scale physical frequency can be calculated by:

$$f_{mel} = 2595 \log_{10}\left(1 + \frac{f}{700}\right) \tag{2.5}$$

where f is the physical frequency in Hz and f_{mel} is the perceived frequency [31]. Later the energy in each filter will be summed, and the logarithm of filter-bank energies will be taken. Finally, discrete cosine transform (DCT) will be applied to the log filter-bank energies, but 12–13 DCT coefficients will be used, and remaining shall be discarded.

2.4.4   Greenwood Function Cepstral Coefficients (GFCC)

The GFCC is a generalised form of MFCC. This feature may provide better vocal representation of many vertebrates since it can be implemented using the information of maximum and minimum frequency range for a species which is derived using the Greenwood equation. This equation tries to map the cochlear frequency position for species. Thus, it is used by many researchers for environmental sound recognition related application, especially in the classification of animals and birds [33]. Before extracting the GFCC features, the audio signal should be framed via suitable windowing techniques, and the periodogram estimate of the power spectrum shall be calculated for each frame. Then the Greenwood-function scaled filter bank is applied to the power spectrum, and later the energy of each filter will be summed. Additionally, the logarithm of filter-bank energies are taken before applying DCT of the log filter-bank energies to produce GFCCs [31]. Greenwood equation can be defined as follows:

$$f = \int_0^x \Delta f_{gf} = S\left(10^{ax} - K\right) \qquad (2.6)$$

where f is the frequency in Hz, S is the scaling constant between the frequencies and maximum frequency limit of the species, K is the constant of integration, x is the functional length and a is the slope of the frequency position curve.

2.4.5   Gammatone Cepstral Coefficients (GTCC)

The GTCC is based on gammatone filter banks, which are generally known to be noise robustness filters. The feature extraction procedure of GTCC is very similar to the extraction of MFCCs, where the Mel filter bank is replaced by a gammatone filter bank. The GTCC can be computed by:

$$f_{gama} = 25.17\left(\frac{4.37 f_c}{1000} + 1\right) \qquad (2.7)$$

where f_c is the physical frequency in Hz and f_{gama} is perceived frequency. From each frame, the power spectrum will be calculated, and gammatone filter bank will be applied. Then the energy of each filter is summed before applying logarithm for each filter-bank energy. Finally, DCT of the log filter-bank energies are taken to produce GTCC features.

2.4.6 Power Normalised Cepstral Coefficient (PNCC)

The PNCC features use the gammatone as the filter bank and apply power-law non-linearity based on the hearing of Steven's power law. It performs a peak power normalisation using the medium-time power bias subtraction method. This will remove the power bias which has information about the background noise level as assumed and uses the ratio of the arithmetic mean to the geometric mean when determining the power bias [34].

2.4.7 Relative Spectral PLP (RASTA-PLP) Feature

The RASTA-PLP is a hybrid feature which is more useful to noisy audio signals, where a bandpass filter can be used to filter the energy in each frequency sub-band. It may help to smooth over short-term noise variations and to remove any constant offset and provide static spectral colouration of the signal. The RASTA-PLP feature usually tries to mimic noise cancellation features of the human auditory system. This feature is commonly employed by a few human speech recognition, speaker verification and gender classification related to vocalisation applications. The pre-emphasis and windowing on the vocal signal are the initial step, where the DFT will be the best option. Subsequently, critical bank analysis on DFT signals will be performed, and then logarithm will be considered, before applying the RASTA filtering. Hence, it may be required to perform the RASTA filtered signals with equal loudness and pre-emphasis. Additionally, intensity loudness power law is applied, and the inverse of the logarithm can be taken to perform auto-regressive modelling. Finally, it can be converted into the cepstral domain to get RASTA-PLP coefficients [31].

2.5 Acoustic Feature Engineering and Feature Selection

In this modern digital era, due to the rapid development of data acquisition and storage techniques, handling of massive data becomes an important process. Even though the vertebrate vocal sound collection is a tedious process, thus developing reliable applications always demands more data. Furthermore, the feature extraction process is vital since the extracted features should be able to provide more information on raw data which impacts the outcome of the applications. In practical scenarios, extracted features may contain some noisy data which are irrelevant and redundant. For example, the extracted features may highly be correlated with one another and give the same pieces of information about the raw data, and some features may not give any optimal enough information for the desired objective of the application, yet few may not have any statistical relationship with the targeted outputs. These types of features may affect the performance of the applications as well as take huge time to process.

Feature engineering and feature selection are the two techniques which are used widely to manage the large scale of extracted features before incorporating them into applications. Even though both techniques have some overlapping characteristics, it is crucial to understand the difference between both techniques. It may help to understand the workflow as well as the pipelines of the application development process. Feature engineering can build more complex features or interpretable features from raw data, where feature selection provides techniques to limit or choose the features based on their significance to the desired goal with manageable numbers. The performance of the machine learning techniques heavily relies on feature engineering and feature selection. It includes dimensionality reduction to extract the feature's value and to avoid the unwanted cost, predictive errors and overlearning. It also focuses on building new feature sets by overlooking machine learning techniques.

Feature engineering allows transforming one or more features into a different representation to provide more impactful information to the model. Thus, it may help to improve the performance of the models by providing a proper input dataset which is compatible with the model requirements. Imputation, handling outliers, binning, log transform, one-hot encoding, grouping operations and scaling are the common feature engineering techniques used by the professionals to improve the features [35].

Feature selection is the widely used technique in modern applications since it can be used to reduce the computation cost and improve the performance of the applications by tackling irrelevant and redundant data issues. It can also be called as variable or attribute selection. This technique also can be used to select the most optimal features for further processing. Generally, it is a process of choosing a feature subset from the original pool of features. The feature selection process can be carried out either manually or automatically based on the size of the feature set as well as the objectives of the application, as an example in most of the supervised machine learning applications to handle the high dimensionality of the data; feature selection can be made prior to the training process [36]. Principal component analysis (PCA) and linear discriminant analysis (LDA) are the most famous dimensionality reduction techniques used as supervised and unsupervised manner, respectively.

Univariate feature selection and multivariate feature selection are the two broad categories of the feature selection process. Univariate feature selection involves manual processes, where the unwanted features are removed by analysing all the features one by one with domain knowledge, but it is a very time-consuming process. There are several tricks which are used to carry out this tedious process, that is, checking the variance, using Pearson correlation. However, it is very complex and challenging to use these techniques when considering a large number of features. Therefore, multivariate feature selection enables the selection of multiple features at once. The ANOVA test, Chi-square test, Lasso regularization in linear regression, select k-best in random forest and gradient boosting machine (GBM) are a few examples of multivariate feature selection methods which can be categorised into three types, namely, wrapping methods, filter methods and embedded methods [37].

2.6 Acoustic Feature Optimisation of Vertebrate Vocalisation Applications

Acoustic analysis and optimisation are useful tools in varieties of applications that are related to the vertebrate vocalisation. Hence, feature optimisation has its own ability to differentiate all kinds of features that are extracted in different domains. Figure 2.4 illustrates the general feature optimisation procedure, which is commonly used in many vertebrate vocalisation related applications.

The general acoustic feature optimisation process, which is shown in Fig. 2.4, involves a few important steps such as input step, where the vertebrate vocal sounds are sent to the designed optimisation process. At the next step, the input vocal signal must be de-noised via suitable filtering techniques. As in the final step, the suitable acoustic features are extracted from the denoised signal. The extracted features can represent the most important information of the signal, as it significantly affects the efficiency of each vertebrate vocalisation related application.

2.6.1 Vocal Signal Input

Input signals of the vertebrates must be in contrast with the common architecture of the vocal signal processing flow. The foremost element of the vertebrate vocal signal flow is to find the sound source, which can be defined as the different types of vertebrates that can be found in nature, that is, amphibians, fishes, reptiles, birds and mammals. Vertebrates produce numerous sounds such as call and speech through their vocal codes. However, the vocal sound capturing process of vertebrates is one of the highly risky and challengeable tasks, since many vertebrates are found in the environments that could be hard to reach by the man. Besides, most of the wild vertebrates are known to be aggressive towards the human. Astonishingly, few people have taken these challenges into their accounts to capture the vocal sounds of vertebrates. Some of them have created online available databases to distribute the recorded vocal sounds to the community.

The 'find sounds' [38] are one of the well-known online databases which includes many vertebrate vocal sounds. The most significant of this database is its fine-tuned

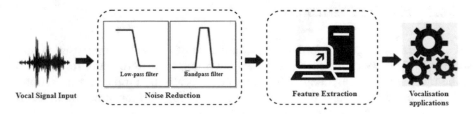

Fig. 2.4 General feature optimisation process for vertebrate vocalisation applications

sound signals, which may not need to be further pre-processed to remove the high-frequency noises. This database is specially invented for the biotechnology and biological science studies that can be integrated with a vast number of bioacoustic-based applications, such as sound classification, hybrid learning environments and active and passive learning [39]. Some of the vertebrate vocal signals are gathered from this database to obtain the performance of the general acoustic feature optimisation process, which is shown in Fig. 2.4. Specifically, due to the processing limitations, only one vocal sample of each vertebrate class including grey tree frog from amphibian class, silver perch from fish class, crocodile from reptile class, masked tityra from bird class and elephant from mammalian class is selected to execute the feature optimisation process.

2.6.2   Noise Reduction

Signal noise reduction takes a major concern in audio signal processing applications because it can better arrange the useful data into proper order before analysis. The main purpose of noise reduction is to remove the effects of the high-frequency noise, which have mixed with the actual sounds while recording. Humans produce the most common high-frequency sounds that are available in the ecosystem. These sounds include traffic, machinery and electronic sounds. However, if the vocal sounds are recorded inside a dense jungle, then there shall be other sounds that may be produced by the trees, wind, water or different species, that is, chirping birds. The low-pass and bandpass filtering processes may be the best solution to remove these types of high-frequency noises. Prior to the signal denoising, these filters convert the signal into the frequency domain and set a frequency band, that is, a low-frequency band to a low-pass filter and both low- and high-frequency bands to a bandpass filter, to analyse the amplitudes of all signal components. Then the frequencies which shall not fulfil by the set frequency bands must be eliminated. These eliminated frequencies are considered as the high-frequency noises of the input signal. Nevertheless, these types of noise can contain large amounts of unwanted data which needs to be separated from the recorded audio signals to extract the useful information for further feature extraction and signal processing tasks.

2.6.3   Feature Extraction

Feature extraction can improve the performance of the feature optimisation process. The quality of the features is very important to acquire better results from their applications, especially in vertebrate vocalisation related applications. Feature extraction stage of the general feature optimisation process, which is shown in Fig. 2.4, shall require the denoising signals to extract the required acoustic features. Among many of these features, time-domain, frequency-domain, and cepstral features are commonly used in audio signal processing applications, including vertebrae vocalisation-based applications. However, the extracted features are represented in a high dimensional space. Thus, suitable feature engineering and feature selection methods shall need to be integrated to dimension reduction and investigate the most accurate features from the extracted features. Nevertheless, the integration of feature engineering and feature selection with the extracted features shall highlight the potential characteristics of the features in order to provide better outputs from many vertebrate vocalisations related applications. In this study, the effectiveness of these features in the acoustic feature optimisation process is investigated via extracting three types of features such as cepstral features, time-related features and frequency-related features. The extracted acoustic features are shown in Table 2.1.

2.6.4   Vocalisation Applications

The acoustic-based applications such as sound classification, emotional recognition, speech recognition and animal behaviour identification are commonly integrated with vertebrate vocalisation. Thus, application-oriented vertebrate vocalisation is the newest trend of creating resistance for climate change, species and biodiversity conservation. These types of applications are commonly found in almost all the fields, including the most devoted fields such as the medical sector, farming and tourism. The acoustic feature optimisation process may have a massive impact on these fields by creating various directions to several new applications and also to develop the existing applications. The applications that relate to vertebrate vocalisation is broadly discussed in Chap. 3 by proposing numerous ways to develop the existing application. Also, Chap. 3 further discusses some of the newest and trending applications which can be possibly implemented as future directions on vertebrate vocalisation-based research works.

Table 2.1 Time-domain, frequency-domain, and cepstral features extracted from the vertebrate vocal sounds

Feature type	Extracted feature
Cepstral	Mel-frequency cepstral coefficients (MFCC), linear prediction cepstral coefficients (LPCC), gammatone cepstral coefficients (GTCC), power normalised cepstral coefficients (PNCC)
Time-domain	Zero-crossing rate (ZCR), energy (E), entropy of energy (EE)
Frequency-domain	Spectral centroid (SC), spectral spread (SS), spectral entropy (SE), spectral flux (SF), spectral roll-off (SR)

2.7   Feature Optimisation Results and Discussions

The significance of the extracted features shown in Table 2.1 was obtained from the analysis of variance (ANOVA) one-way test by keeping one independent feature, whilst other extracted features are considered as dependent features. The ANOVA test allows to find out if the null hypothesis can be accepted or rejected in order to analyse the statistical significance and strength of the extracted features of different classes [40]. Moreover, this test uses F-test as a part of its statistical analysis to investigate the equality of the means of each feature group.

F-test considers the ratios of two variances to identify how far the data is distributed from the mean. However, the square of standard deviation is called variance. Therefore, F-statistical analysing process is generally based on the mean squares of the ratios, but in ANOVA analysis, the mean square is estimated through the population variance which is named as the degree of freedom (DF) [41]. Nevertheless, F-test can be applied in various applications which includes audio processing applications to get better results in the acoustic feature analysing tasks.

2.7.1   Cepstral Feature Analysis

The cepstral features, that is, MFCC, LPCC, GTCC and PNCC, are extracted for 13 coefficients, yet all these features are in a high dimension, which may make it hard to identify which coefficient shall provide a high impact in the vertebrate vocalisation related applications. Therefore, the linear discriminant analysis (LDA) feature dimension reduction method is applied to all four types of extracted cepstral features. The LDA is a simple linear feature processing method that allows identifying the best features from all features. It sends the best feature combinations to the front columns of the feature matrix. In other words, the most significant and highly impactful feature will be set in the first column of the feature matrix. Therefore, the use of this method helped to determine the best coefficient from all selected cepstral features. Then the ANOVA one-way test is performed to the selected coefficient values of all four types of cepstral features, but there were total two experiments for these features: (1) considering MFCC as an independent feature along with GTCC and PNCC as dependent features and (2) considering LPCC as an independent feature along with GTCC and PNCC as dependent features. Hence, MFCC and LPCC are commonly found in the past research studies, and they are considered to provide better results for many vertebrate vocalisation related applications [3, 5, 7, 8, 14, 42]. The two ANOVA one-way test results of selected cepstral features are shown in Table 2.2.

As shown in Table 2.2, it can be seen that both (1) and (2) experiment results are equal. Most significantly, their P-values (2.515×10^{-16}) are very small compared to the preset significance level (α) value of 0.05, which allows rejecting the null hypothesis (H0). This may happen due to the low variability, which tends to keep class (all four feature types) mean values closer to each other while keeping

Table 2.2 ANOVA test results for selected cepstral features

Experiment	Source	DF	SS	MS	F-value	P-value
1. MFCC as an independent feature along with GTCC and PNCC as dependent features.	Factor	01	1.3205	1.3205	45929.6	2.515 $\times 10^{-16}$
	Error	08	0.0002	0.0000287		
	Corrected total	09	1.3207			
2. LPCC as an independent feature along with GTCC and PNCC as dependent features.	Factor	01	1.321	1.321	45929.6	2.515 $\times 10^{-16}$
	Error	08	0.0002	0.0000287		
	Corrected total	09	1.321			

SS sum of squares, *MS* mean of squares, *DF* degree of freedom

proportional variability on each class. In addition, the high F-value (45929.6) also took part in rejecting the H0. Thus, F-value is computed by the mean squares (MS) of factor dividing with the MS of error. Moreover, it shows that the obtained results are consistent with verifying the impact of both MFCC and LPCC features in vertebrate vocalisation analysis applications, since these two cepstral features gave better feature analysis results, separately compared to GTCC and PNCC features.

2.7.2   Time-Domain and Cepstral Feature Analysis

The ANOVA one-way test justification of the MFCC and LPCC feature contribution to vertebrate vocalisation applications is considered to compare the effect of the extracted time domain features, which is shown in Table.2.1. These time-domain features are compared with both MFCC and LPCC cepstral features as two separate ANOVA one-way experiments, that is, (3) MFCC as an independent feature along with time-domain features as dependent features and (4) LPCC as an independent feature along with time-domain features as dependent features. The obtained ANOVA one-way test results are shown in Table 2.3.

The ANOVA one-way test results of both experiment (3) and (4) provide almost similar results. Although due to the calculated p-value (4.027×10^{-11}) that is less than the α value of 0.05, both experiments successfully rejected H0. Moreover, the results in Table 2.3 prove that both MFCC and LPCC features can contribute well in several vertebrate vocalisation applications compared to the time-domain features such as zero-crossing rate (ZCR), energy (E) and entropy of Energy (EE).

2.7.3   Frequency-Domain and Cepstral Feature Analysis

A similar type of ANOVA one-way tests is performed to obtain the efficiency of the MFCC and LPCC features compared to the frequency-domain features shown in Table 2.1. Therefore, another two separate experiments, that is, (5) MFCC as an independent feature along with frequency-domain features as dependent features

Table 2.3 ANOVA test results for time-domain, MFCC, and LPCC features

Experiment	Source	DF	SS	MS	F-value	P-value
3. MFCC as an independent feature along with time-domain as dependent features.	Factor	02	18.373	9.186	318.07	4.027 ×10⁻¹¹
	Error	12	0.346	0.0288		
	Corrected total	14	18.719			
4. LPCC as an independent feature along with time-domain as dependent features.	Factor	02	18.373	9.187	318.075	4.027 ×10⁻¹¹
	Error	12	0.347	0.029		
	Corrected total	14	18.720			

SS sum of squares, *MS* mean of squares, *DF* degree of freedom

Table 2.4 ANOVA test results for frequency-domain, MFCC and LPCC features

Experiment	Source	DF	SS	MS	F-value	P-value
5. MFCC as an independent feature along with frequency-domain as dependent features.	Factor	04	4.5097	1.12743	9.584	0.000169
	Error	20	2.3525	0.11762		
	Corrected total	24	6.8622			
6. LPCC as an independent feature along with frequency-domain as dependent features.	Factor	04	4.5097	1.12743	9.584	0.000169
	Error	20	2.3525	0.11762		
	Corrected total	24	6.8622			

SS sum of squares, *MS* mean of squares, *DF* degree of freedom

and (6) LPCC as an independent feature along with frequency-domain features as dependent features, have been done to investigate how the extracted frequency-domain features may affect to the vertebrate vocalisation related applications. The experiment results are shown in Table 2.4.

The results of Table 2.4 show that both MFCC and LPCC independent features with frequency-domain dependent features have managed to reject the H0 by keeping a low p-value than the set α value of 0.05. Thus, the selected independent features are comparably better than the extracted frequency-domain features such as spectral centroid (SC), spectral spread (SS), spectral entropy (SE), spectral flux (SF) and spectral roll-off (SR). However, it can be noted that the F-value of experiments (5) and (6) are a lot smaller than the other four experiments, but the total number of observations (corrected total) of experiments (5) and (6) are larger than the other four experiments. Hence, this may be caused due to the selected number of features; thus experiments (5) and (6) contain total seven features including both independent and dependent, while experiments (1) and (2) are contained with four features, and experiments (3) and (4) are contained with five (05) features.

Moreover, it can be noted that the cepstral features have considerable influence to get better statistical analysis results. Specifically, the MFCC and LPCC features are much better than GTCC and PNCC features. Besides, the MFCC and LPCC features also perform better than the extracted time- and frequency-domain features. However, the ANOVA one-way test results obtained in this section may not be more accurate for any other vertebrate vocal sounds, yet it still can utter all the features better with their specific application. Nevertheless, the selected number of vocal recordings has been limited to one for each vertebrate class such as amphibians, fishes, reptiles, birds and mammals. These limitations are concerned in this section only to observe the effect of these features for the selected vocal recordings. Thus, these results may be changed for other vocal recordings. Therefore, these results are considered to check the reliability and efficiency of each feature type when applying for vertebrate vocalisation related applications. In addition, these results shall be improved to get more accurate information and outcomes for all the vertebrate vocalisation related applications by increasing the number of observations as well as machine learning techniques.

2.8   Summary

The acoustic features are important to be extracted from vertebrate vocal signals to achieve the most accurate results from the related applications that are applied in various fields such as medical, industrial, biological and many more. Among many types of acoustic features, the time-domain, frequency-domain and cepstral domain features have provided more consistent outcomes in vertebrate vocalisation related applications. Specifically, homomorphic signal processing has taken major popularity in deriving acoustic features that can be applied in numerous signal processing applications, including speech recognition and image enhancement. Subsequently, homomorphic signal processing has its unique facility to enhance the speech analysis along with signal echo removal. Nevertheless, cepstrum of a signal is also referred to as homomorphic signal processing, which tends to extract the cepstral features such as MFCC, LPCC, GFCC, GTCC and PNCC. The cepstral features have strongly influenced the vertebrate vocal signal processing applications. Thus, there are many available vertebrate vocalisation applications which are based on these cepstral features, especially MFCC and LPCC features.

The acoustic feature engineering and best feature selection are considered as the key success of the outcome of audio signal processing applications, specifically voice related applications. Both feature engineering and feature selection are considered as separate techniques, where feature engineering builds complex features while feature selection is identifying the significance of these complex features. Nevertheless, both techniques aid to improve the performance of outcomes of their applications by inputting the best features which are compatible with the application requirements.

Moreover, the feature engineering and feature selection processes have been implemented in this chapter to investigate the most suitable features for vertebrate vocalisation applications. However, a few vocal signals are considered to perform the selected feature testing experiments due to the experimental limitations, but one vocal sample from each vertebrate category, that is, amphibians, fishes, reptiles, birds and mammals, is considered. Prior to extracting the specified time-domain, frequency-domain and cepstral features, the audio signals are denoised using suitable low-pass and bandpass filters. Then the required features such as spectral centroid (SC), spectral spread (SS), spectral entropy (SE), spectral flux (SF), spectral roll-off (SR), zero-crossing rate (ZCR), energy (E), entropy of energy (EE), MFCC, LPCC, GTCC and PNCC features are extracted. Before applying these extracted features to the ANOVA one-way feature testing process, the linear discriminant analysis (LDA) has applied only to the extracted cepstral features such as MFCC, LPCC, GTCC and PNCC to investigate the best feature from all the extracted features, separately.

The ANOVA one-way test has conducted as six experiments, namely, (1) considering MFCC as an independent feature along with GTCC and PNCC as dependent features, (2) considering LPCC as an independent feature along with GTCC and PNCC as dependent features, (3) MFCC as independent features along with time-domain features as dependent features, (4) LPCC as independent features along with time-domain features as dependent features, (5) MFCC as independent features along with frequency-domain features as dependent features, and (6) LPCC as independent features along with frequency-domain features as dependent features. The experiments number (1) and (2) have proved the MFCC and LPCC feature may have a considerable impact on the selected vocal readings. Nevertheless, this has proved more when testing the MFCC and LPCC as dependent features to both outstanding time domain and frequency domain (as in experiment numbers (3), (4), (5) and (6)). Therefore, all performed ANOVA one-way tests have proved cepstral features may have a strong influence to the selected vertebrate vocal signals, which can be further elaborated to define as these features can be statistically proven to be used in almost all the vertebrate vocalisation related applications.

References

1. S. Shamma, The acoustic features of speech sounds in a model of auditory processing: Vowels and voiceless fricatives. J. Phon. **16**(1), 77–91 (1988)
2. J.S.J. Wolfe, *Voice Acoustics: An Introduction* (Acoustics Group at UNSW, 2013) [Online]. Available: https://newt.phys.unsw.edu.au/jw/voice.html
3. M.P. Mcloughlin, R. Stewart, A.G. McElligott, Automated bioacoustics: Methods in ecology and conservation and their potential for animal welfare monitoring. J. R. Soc. Interface **16**(155) (2019)
4. K.R. Paap, Theories of speech perception, in *Understanding Language*, (Elsevier, 1975), pp. 151–204
5. J.H.L. Hansen, M.K. Nandwana, N. Shokouhi, Analysis of human scream and its impact on text-independent speaker verification. J. Acoust. Soc. Am. **141**(4), 2957–2967 (2017)
6. P.J. Clemins, M.T. Johnson, K.M. Leong, A. Savage, Automatic classification and speaker identification of African elephant (Loxodonta africana) vocalisations. J. Acoust. Soc. Am. **117**(2), 956–963 (2005)
7. P.J. Venter, J.J. Hanekom, Automatic detection of African elephant (Loxodonta africana) infrasonic vocalisations from recordings. Biosyst. Eng. **106**(3), 286–294 (2010)

8. M.J. Owren, R.M. Seyfarth, D.L. Cheney, The acoustic features of vowel-like grunt calls in chacma baboons (Papio cyncephalus ursinus): Implications for production processes and functions. J. Acoust. Soc. Am. **101**(5), 2951–2963 (1997)
9. D.A. Mann, Remote sensing of fish using passive acoustic monitoring. Acoust. Today **8**(3), 8 (2012)
10. M.A. Graham, M. Noonan, Call types and acoustic features associated with aggressive chase in the killer whale (Orcinus orca). Aquat. Mamm. **36**(1), 9–18 (2010)
11. L. Nanni et al., Bird and whale species identification using sound images. IET Comput. Vis. **12**(2), 178–184 (2018)
12. Z.D. Burkett, N.F. Day, O. Peñagarikano, D.H. Geschwind, S.A. White, VoICE: A semi-automated pipeline for standardising vocal analysis across models. Sci. Rep. **5**, 1–15 (2015)
13. M. Ramashini, P.E. Abas, U. Grafe, L.C. De Silva, Bird sounds classification using linear discriminant analysis. ICRAIE 2019 – 4th Int. Conf. Work. Recent Adv. Innov. Eng. Thriving Technol. **2019**(November), 27–29 (2019)
14. D. Hosseinzadeh, S. Krishnan, Combining vocal source and MFCC features for enhanced speaker recognition performance using GMMs. 2007 IEEE 9Th Int. Work. Multimed. Signal Process. MMSP 2007 – Proc., 365–368 (2007)
15. D. Rossiter, D.M. Howard, M. Downes, A real-time LPC-based vocal tract area display for voice development. J. Voice **8**(4), 314–319 (1994)
16. N. Dave, Feature extraction methods LPC, PLP and MFCC in speech recognition. Int. J. Adv. Res. Eng. Technol. **1**(Vi), 1–5 (2013)
17. A.L. McIlraith, H.C. Card, Birdsong recognition with DSP and neural networks. IEEE WESCANEX Commun. Power, Comput. **2**(95), 409–414 (1995)
18. D. Wang, L. Zhang, Z. Lu, K. Xu, Large-scale whale call classification using deep convolutional neural network architectures. 2018 IEEE Int. Conf. Signal Process. Commun. Comput. ICSPCC, 1–5 (2018)
19. T. Zhang, Y. Shao, Y. Wu, Z. Pang, G. Liu, Multiple vowels repair based on pitch extraction and line Spectrum pair feature for voice disorder. IEEE J. Biomed. Heal. Inform. **24**(7), 1940–1951 (2020)
20. S.H. Fang et al., Detection of pathological voice using Cepstrum vectors: A deep learning approach. J. Voice **33**(5), 634 641 (2019)
21. G. Jhawar, P. Nagraj, P. Mahalakshmi, Speech disorder recognition using MFCC. Int. Conf. Commun. Signal Process. ICCSP **2016**, 246–250 (2016)
22. I. Potamitis, S. Ntalampiras, O. Jahn, K. Riede, Automatic bird sound detection in long real-field recordings: Applications and tools. Appl. Acoust. **80**, 1–9 (2014)
23. S. Nowicki, P. Marler, *How Do Birds Sing?* (Music Percept. An Interdiscip. J, 1988)
24. D.J. Jin, E. Eisner, A review of homomorphic deconvolution. Rev. Geophys. **22**(3), 255–263 (1984)
25. L. Su, Between homomorphic signal processing and deep neural networks: Constructing deep algorithms for polyphonic music transcription. Proc. – 9th Asia-Pacific Signal Inf. Process. Assoc. Annu. Summit Conf. APSIPA ASC 2017 **2018**, 884–891 (2018)
26. X. Zhao, D. Wang, Analysing noise robustness of MFCC and GFCC features in speaker identification, in *ICASSP, IEEE International Conference on Acoustics, Speech and Signal Processing – Proceedings*, (2013), pp. 7204–7208
27. J.R. Deller, J.H.L. Hansen, J.G. Proakis, *Discrete-Time Processing of Speech Signals* (2010)
28. L.R. Rabiner, B. Gold, Theory and application of digital signal processing. IEEE Trans. Acoust. Speech Signal Process. **23**(4), 394–395 (1975)
29. D.G. Childers, D.P. Skinner, R.C. Kemerait, The Cepstrum: A guide to processing. Proc. IEEE **65**(10), 1428–1443 (1977)
30. V.R. Reddy, *Language Identification Using Spectral and Prosodic Features* (Springer, 2015)
31. G. Sharma, K. Umapathy, S. Krishnan, Trends in audio signal feature extraction methods. Appl. Acoust. **158**, 107020 (2020)

32. A.G.F. Properties, Gammatone cepstral coefficients: Biologically inspired features for non-speech audio classification. **14**(6), 1684–1689 (2012)

33. K. Adi, M.T. Johnson, T.S. Osiejuk, Acoustic censusing using automatic vocalisation classification and identity recognition. J. Acoust. Soc. Am. **127**(2), 874–883 (2010)

34. A. Badi, K. Ko, H. Ko, Bird sounds classification by combining PNCC and robust Mel-log filter bank features. J. Acoust. Soc. Korea **38**(1), 39–46 (2019)

35. Ananyd, *7 Feature Engineering Techniques in Machine Learning you Should Know* (Analytics Vidya, 2020) [Online]. Available: https://www.analyticsvidhya.com/blog/2020/10/7-feature-engineering-techniques-machine-learning/

36. D.A.A. Gnana, Literature review on feature selection methods for high-dimensional data. **136**(1), 9–17 (2016)

37. A. Desarda, *Getting Data ready for modelling: Feature engineering, Feature Selection, Dimension Reduction*, towardsdatascience.com, 2018. [Online]. Available: https://towardsdatascience.com/getting-data-ready-for-modelling-feature-engineering-feature-selection-dimension-reduction-39dfa267b95a

38. "FindSounds." [Online]. Available: https://www.findsounds.com/ISAPI/search.dll

39. X. Z. Wenjing Han, Eduardo Coutinho, Huabin Ruan, Haifeng Li, Björn Schuller, Xiaojie Yu, "Dataset," 2016. [Online]. Available: https://plos.figshare.com/articles/dataset/Description_of_the_subset_of_the_FindSounds_database_used_in_this_paper_/3832497/1

40. S. Glen, *ANOVA Test: Definition, Types, Examples*, StatisticsHowTo.com: Elementary Statistics for the rest of us! [Online]. Available: https://www.statisticshowto.com/probability-and-statistics/hypothesis-testing/anova/#top

41. Understanding Analysis of Variance (ANOVA) and the F-test, Minitab blog (2016). [Online]. Available: https://blog.minitab.com/blog/adventures-in-statistics-2/understanding-analysis-of-variance-anova-and-the-f-test

42. A.C. Eva Kiktova, M. Lojka, M. Pleva, J. Juhar, Comparison of different feature types for acoustic event detection system. Commun. Comput. Inf. Sci. **368**. CCIS(June) (2013)

Chapter 3
Trending Technologies in Vertebrate Vocalisation Applications

3.1 Introduction

The vertebrate vocalisation takes a significant role in many habitual activities, and especially wild vertebrates are one of the key resources that can highly impact the human habitant. Vertebrate communication can be named as the most devoted area of study in many research fields including medical, biology, zoology and farming. In the past, people have recorded various types of animal, insect and bird sounds like a hobby, and later these sounds were used to analyse the vocalisation patterns to identify the types of animals and their breeds. Thus, audio signal processing was popular in history before the development of advanced photography and image processing techniques. The growth of research studies and advanced technologies has motivated many people to gather the sounds of vertebrates, mainly targeting commercial usage. As a result, there are various types of devices, that is, underwater listening devices and advanced microphones, that have been improved to record the sounds of nature.

Nowadays, many advanced audio processing devices are used in numerous applications, that is, animal behaviour training, emotion recognition, speech processing, language translation, animal and bird watching, teaching tools, medical diagnosis, biological and biomedical research studies and farming. Integration of the trending technologies such as artificial intelligence (AI), augmented reality and virtual reality (ARVR), cognitive computing, cloud computing, DevOps, internet of things (IoT), intelligent apps (I-Apps), big data and robotic process automation (RPA) with the acoustic features can be worthwhile to humankind to address the future development of the world's needs. Nevertheless, the proper way of using these technologies can be beneficial to reduce the effect of the world's top-level threats such as global warming and climate change. Consequently, it will help to protect the ecosystem and its biodiversity.

© The Author(s), under exclusive license to Springer Nature   39
Switzerland AG 2022
R. Murugaiya et al., *Acoustic-Based Applications for Vertebrate Vocalization*,
SpringerBriefs in Applied Sciences and Technology,
https://doi.org/10.1007/978-3-030-85773-8_3

3.2   Significant of Technology Integration for Vertebrate Vocalisation Related Applications

The needs of technological integration for vertebrate vocalisation related applications are numerous. It helps to identify the damages in the natural ecosystem created by human activities and provide excellent solutions for understanding the species biodiversity and the climatic changes in the ecosystem. Some of these needs are summarised in Fig. 3.1.

Explicit knowledge on ecosystems including population density, habitat loss and environmental pollution can be obtained by monitoring its species biodiversity. Technologies associated with bioacoustic monitoring are helpful if the endeavour of data collection and processing is relatively more straightforward than the human observance. Data collected from acoustic monitoring are used in determining the variation of sound. At first, the vocal sounds are recorded from species. Then it is extracted to use it for various purposes like identifying species abundance and habitat preferences, identifying juveniles and detecting predator's presence by alarm calls [1]. Some tools like syntactic pattern recognition to detect characteristic 'whip-crack' of *Psophodes olivaceus*, hidden Markov models for the frequency modulated whistle of the *Strepera graculina* and binary template matching that was used to detect the pulsatile bellows of male koalas were used as a bioacoustic monitoring tool. It has found to be convenient and reliable for recognisers. Vocal instability is considered as a challengeable task in monitoring the species under open-ended vocal learning, that is, species having the ability to learn vocalisation throughout their lifetime. For instance, vocal sounds of palm cockatoos vary over time. Therefore, the development of a highly specific recogniser is needed for these kinds of species [2].

Vocalisation plays a vital role in conserving endangered species. In China, a survey was conducted based on vocalisation to conserve Gaoligong hoolock

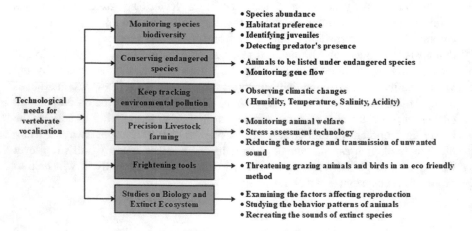

Fig. 3.1 Needs for technological integration with vertebrate vocalisation

gibbon, and also the authors stated that it should be listed under endangered species [3]. A population-specific vocal feature termed as dialect plays a crucial role in a variety of processes in avian conservation including territory formation, mate selection and gene migration. Genetic mixing between populations results in poor gene flow. For instance, inbreeding of red-cockaded woodpecker results in the least hatchling success and fledgeling survival. Thus, for conserving birds, particularly endangered species, technologies cooperated with monitoring the gene flow may help to protect them from destruction [1].

Sound absorption is a phenomenon used in bioacoustics, in which the medium increases its internal energy by absorbing encountered sound waves. Vocal communication of birds and animals gets different according to the seasonal changes because acoustic absorption highly depends on humidity, temperature, pressure, salinity, acidity and other physicochemical parameters. For instance, the echolocation sounds of bats are found to be longer with lower frequency in the rainy season than in other dry seasons due to the absorption of atmospheric humidity and precipitation. Hence, the impacts of environmental pollution can be easily monitored by observing the vocal behaviour of birds. Increased ambient noise and ocean acidification may affect the vocalisation abilities in marine vertebrates, while in terrestrial vertebrates, changes in the environmental temperature and humidity may modify their vocalisation abilities. Both marine and terrestrial vertebrate's vocal sounds get modified due to the changes in the physical parameters of the ecosystem that results in global warming, which leads to glaciers melting and finally ends up with natural calamities. On account of this, many technologies emerged to monitor the environmental changes by collecting the vocal records of vertebrates for protecting the ecosystem [4, 5].

In livestock farming, vocalisation displays a hint to recognise the physical wellness of animals. Therefore, there is a need for emerging new technologies to monitor the health and body conditions. The ultimate goal of livestock farming is to obtain information about animal welfare and to improve production by using software and hardware systems. Livestock management approaches the precision livestock farming that concerns on automated detection and classification of audio signals [6]. Wasting disease is a contagious disease in livestock which requires timely detection [4]. Through monitoring the cough sound produced by the livestock, the disease can be easily identified. Hence, highly efficient technology is needed to capture the particular data in a noisy environment. To avoid the noisy environment in the poultry, an algorithm was developed based on power spectral density to detect the vocalisation of laying hens. It shows the positive response in reducing the storage and transmission of unwanted sound and improves the sound analysis efficiency [7]. In pigsty, respiratory disorders of pigs were diagnosed by analysing the cough sound produced by pigs. Cattle welfare was identified by capturing vocal behaviour, including murmuring sound during their resting period. The flaws in the cow's oestrous cycle were detected by obtaining its vocal sounds using noise sensors to determine the effect of thermal condition on chick's health. These are few findings using technologies in acoustic monitoring for livestock management. Moreover, stress is assumed to have deleterious effects on the production of commercial chickens. Vocalisation is an authentic way to quantify the stress in chickens [5]. Thus, simple, accurate and fastest stress assessment technology would be preferable.

Agricultural producers in rural and suburban areas are adversely affected by birds. Birds create many problems in agricultural lands by feeding and trampling crops and also involve in spreading disease by leaving their droppings. Waterbirds including geese, ducks, cranes and swans pollute water by means of roosting, feeding and loafing, which further result in poor sanitary conditions. Many countries try to keep away the birds from that particular area by deploying visual deterrents like lasers and some techniques including pyrotechnics, gas explorers and ultrasonic devices. Conversely, the pollutants released during those techniques affect the lungs, and it gives a negative impact by altering the behaviour patterns of birds such as reduced hatching productions, thinning of eggshells and changes in vocalisation and cause cancer which results in death. Birds must be conserved because it plays an important role in the ecosystem as a predator, pollinator and scavenger and helps to disperse seeds [8]. Hence, birds must be threatened in an eco-friendly way. Some research revealed that animals and birds react physiologically to the distress calls produced by them, so frightening devices which exhibit distress calls are evolved as integrated pest management to reduce the damages caused by grazing animals and birds [9]. On hearing the recorded sounds of distress calls, adrenalin levels of starlings get increased, and it stimulates the speed of the heart rate. Then they emotionally get disturbed and start to fly away. Birds like Canada geese and crows show similar responses against the distress calls and disperse away from the area.

Moreover, to prevent the endangered species from its predators, tape recordings of the vocalisation are in use. For instance, sugar gliders are small mammals which predate the hatchlings of swift parrots, an endangered species in Australia. In order to prevent swift parrots from sugar gliders, the calls of southern boobook owls are used as an alarm which depredates the sugar gliders [2]. Thus, techniques associated with vocalisation were implied to frighten the birds and some predators to drive away from that region in an environmentally friendly way.

Male vertebrates use ultrasonic vocalisation for mating and courtship. Emotional disturbance or increased anxiety levels may alter the courtship vocalisation and mating. Hence, male sexual behaviour is tightly coupled with neuronal circuits. Adopting this mechanism, new researches were investigated to determine the factors affecting the male reproductive system by recording the vocal sounds and analysing the data under chronic exposure of the male vertebrate to the particular factor as a testing material. For instance, research on adult mice which get exposed to nonylphenol alters its vocal behaviour during courtship and mating. Therefore there is a need for raising new technologies for solving the problems in human reproduction, a major issue faced by many couples in this modern world [10]. Concealed behaviour of some species, especially the elephant's in situ activity patterns, was clearly understood by means of bioacoustics. Thus, technologies associated with vocalisation help to study the animals that are difficult to find.

A knowledge of historical information of the past ecosystem can be obtained by collecting the vocal records of some extinct species such as yellow warbler and Darwin's frog. It was collected as a fossil record, and it helps to gain information about the existing ecosystem. Palaeontologists have tried to recreate the sounds of some extinct creatures like dinosaurs, hominids and crickets. Technologies created using appropriate practices can help to learn about the drastic changes in the ecosystem [5].

3.3 Citizen Science Projects for Vertebrate Vocalisation

Citizen science is a technique involving the public through participating them in scientific research-related activities. Most of the cases it allows people to share and contribute to the data collection or monitoring phase. The overall aim is to provide awareness about scientific research to the public by involving them and improve their science and technological knowledge [11]. Researchers and scientists can build collaborations with community-based groups such as weather bugs, amateur groups and bird watchers to extend the studies and database since they may have already collected data. These projects would allow them to capture more and more widely spread data economically without investing much funding. Interested amateur scientists, students and educators shall volunteer to develop a network and encourage new ideas to advance their knowledge of the world. Expertise level of the volunteers may vary from kids to elders, where kids may collect data from their backyards, school science club to amateur students or researchers with equipment handling knowledge. Citizen science is a perfect platform for students to get practical real-world knowledge and expert's advice before deciding their career path. The project success may heavily depend on how well the devised monitoring program and involvement of its volunteers.

Vertebrate vocalisation collection tasks will flourish with citizen science projects because the outcome of the application can often be correlated with the collected data, and the time that can be spent on the data collection process will be reduced drastically with these projects. It can also be developed as a wildlife monitoring program with proper training and implementation. However, vertebrates are diverse in nature. Therefore, citizen science projects can be implemented as global networks to exchange and collect more information about the vocalisation of vertebrates using digital platforms.

Furthermore, modern advances in technology make citizen science more accessible and lead to expanding it for online databases, visualisation and sharing technologies. As a future direction, citizen science projects can be armed with trending technologies, that is, AI, IoT and cognitive systems, to gather more metadata along with vocal data of varieties of species in real time. This can open up many potential future opportunities to create new inventions to develop vertebrate vocalisation applications.

3.4 Potential Future Directions on Development of Vertebrate Vocalisation Applications

There is a vast spectrum of issues that can be addressed using the trending technologies to authenticate vertebrate vocalisation applications, which is elaborated in Sect. 3.2. However, in the technological world, anything is possible if the proposed concept can be matched with suitable technology. There are a certain number of applications that can provide the most reliable outcomes through these technologies. Apart from all of these applications, the vertebrate vocalisation-based application plays a vital role due to its vast range of useful information. It is important to thoroughly consider the selection of the input data because it is the key success of the modern-day trending technologies.

3.4.1 Artificial Intelligence (AI)

Artificial intelligence (AI) is one of the top-level advanced modern technologies that can combine humans and machines. This often refers to machine intelligence because it is capable of thinking as human and imitating their actions. AI can learn faster than a human brain, and it can provide accurate predictions and solutions for any problematic task. Nowadays, AI has developed as a self-learning tool to improve its decision-making and rationalise skills. AI is a vast area of study that includes both machine learning and deep learning.

Many research studies adopt the vocal information of birds and animals, specifically vertebrates for the reason that the vocal sounds contain extensive information [12–14]. However, it may be problematic to manage the standardised vocal information across a huge geographic area, but this can be avoided using the AI technology to create an automated vertebrate vocal recording system. These recorded vocal sounds can be used to recognise each vertebrate species by creating an automatic vertebrate vocalisation recognition system. This system can be created using a suitable machine learning approach. Furthermore, this automated vocalisation recognition system can be developed as a software that can be used in a personal computer for the reason that it can be used by any user.

Most animals, including vertebrates, can understand human emotions and commands. As a result, some intelligent vertebrates can mimic human voices and actions. There are many ways to measure human intelligence level. Thus, these measurements may not be suitable to measure the intelligence level of other vertebrates. AI technology can provide a great impact to find more reliable measurements to measure the intelligence level of the vertebrates. However, there are a certain number of research studies that approach on vertebrate vocalisation translation using AI [15, 16]. However, these types of application can be developed vice versa to investigate the intelligent level of these vertebrates and how well they can understand different types of human languages. Therefore, it can create a strong bond towards humans and other vertebrates.

Nowadays, human speech recognition is widely popular among people, due to the introduction of smart devices such as phones, computers and tabs. There are certain AI-based speech recognition platforms, that is, Google Assistant and iPhone Siri. In recent years, these platforms have integrated much more novel features to provide high-quality and reliable outcomes. Therefore, they can be used to create intelligent apps (I-Apps) that can convert historical and real-time vertebrate vocalisation data to enable much more advantages in vocal data analytics. This proposed I-App is shown in Fig. 3.2. Hence, the combination of smart concept and AI may help I-Apps to provide numerous output results, that is, types of vertebrates, their livelihood, health status, breed, age and gender. This proposed I-App may help people who cope with animals and birds to know the information that is hidden in their vocals in real time via a smart device.

Another aspect of AI implementation for vertebrate vocalization is forecasting methods. Utilizing historical vertebrate vocalisation data collected over the years can be used to predict the future via AI forecasting methods. It will be an interesting technology for many of applications, that is, endangered species prediction, climatic change tracking and species abundance calculation. Reliability of the predictions in forecasting-based applications will always be correlated with the data used for the model training. Therefore, it is essential to establish a well-defined data collection phase to flourish the benefits of the technology.

3.4.2 Internet of Things (IoT)

Internet of things (IoT) refers to one of the highly influenced technologies that can globally connect billions of physical devices to collect and share data. This technology integrates any objects via embedded systems to create a communication link with humans as well as other devices [17]. IoT technology is popular as a 'research people' concept because they create new concepts and aspects almost every day. In recent years, IoT has introduced many remarkable opportunities to develop a multidisciplinary field of studies, that is, engineering, medical and economical. Thus, the use of IoT has resolved a lot of mysterious practical and technical matters. There are a vast number of advantages of using IoT in many applications, especially it can create novel applications that can provide high-quality outcomes. Out of many other applications, IoT can provide utmost benefits in acoustic feature analysis.

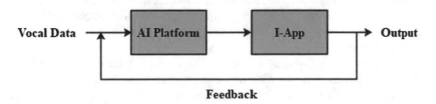

Fig. 3.2 I-App for vertebrate vocalisation recognition

Subsequently, acoustic features are known to be highly accurate features that can provide a massive range of information.

The use of IoT in vertebrate vocalisation can be beneficial in many ways to expand and open up numerous and unique research categories. IoT can recommend the ability to upgrade various sound listening devices to predict the sounds of vertebrates. Handy devices, that is, wristwatch and a mobile app, can be developed using IoT for animal watchers, bird watchers, campers, hikers and ecologists. These devices can be used to identify the breed and location of vertebrates, in addition as a safety device or learning material. For example, the user can identify the location of any dangerous animals ahead. Hence, IoT can be used to connect one device with another to use as a device location identifier, that is, if someone is lost inside a jungle, the location tracker would track the sound of the vertebrate that may be near to him, then he can use this technology to share his location with the rescue team. Moreover, this device can be developed to predict the vertebrate's health condition, emotions and physical behaviours. Several recipes, that is, generating reports, vertebrate growth prediction, ovulation prediction, migration and behaviour analysis, can be implemented to this device to get better advantages. This device may be beneficial to any user, especially to vets and animal science researchers. In addition to this device, IoT can be useful to build smart devices such as biochips, wild vertebrate counting devices and many more. The information that can be gathered from these types of devices can be useful for the present and future vertebrate vocalisation-based research studies to provide feasible solutions for many unsolved problems.

The concept of IoT for vertebrate's vocalisation can be further elaborated to predict the needs, that is, hunger and the thirst of domestic vertebrates. They specifically require higher attention from their owners. Compared to wild vertebrates, domestic vertebrates have strong communication links with humans. Thus, they are good at expressing their feelings to the human with various vocal sounds such as whines, barks and meows. Creation of a mobile app using these sounds can be highly beneficial to the people who work on busy schedules in many ways, especially to feed their loving pets on time.

Vertebrates share various physical and emotional similarities with each other. They are named as the most socialist and intelligent creatures in the animal kingdom. They are eligible to communicate with their community and create a strong bond to protect each other from any threat. However, the human to other vertebrate communication is still lacking due to many ruthless activities of the human towards them. Nevertheless, IoT can be used as a handling tool to create a strong communication bond between the human and the other vertebrates. This can be achieved by the IoT-based cameras that have microphones and speakers, which allow the human to see and talk to the other vertebrates. Furthermore, this technology can be worthwhile to the pet owners to be aware of communicating with their pets while they are away from home.

Natural disaster management is another crucial application that can use IoT, yet there is no such natural disaster identification system using the vertebrate vocalisation. Many types of vertebrates have highly accurate sensing systems to predict any natural disaster beforehand. However, there are a certain number of vertebrates that can alarm these disasters through vocalisation. Especially, the birds are very sensitive to the changes in the air pressure. Thus, they release high-pitched calls to alarm the same species about the upcoming natural disasters, that is, storms and tornados. The sea vertebrates such as dolphins are known as one of the friendly vertebrates to the human; thus, they are well known as a navigator and survivor to the underwater divers and fishers. However, dolphins can predict natural disasters such as tsunamis, and earthquakes before the human do. Therefore, the alarming calls of the vertebrates can be a useful input to create a natural disaster management system that can apply in the land, air and water. These calls can integrate with IoT to alarm the upcoming natural disasters globally to take early precautions to prevent severe damages. This application will be beneficial to any user, mostly fishers, sailors, pilots, underwater divers and archaeologists.

Apart from the advantages of IoT for vertebrate vocalisation, few more disadvantages need to be addressed. The main disadvantage of using IoT is its requirement of a strong internet connection. Therefore, almost all of the proposed applications shall be used within an area that has internet access. However, this needs to take into consideration to obtain more accurate information and create revolutionary changes in vertebrate's livelihood.

3.4.3 Augmented Reality (AR)

Augmented reality (AR) converts digital elements to a live view using an electronic device, that is, a camera on a smartphone. Most popular AR experiences are the 'Pokemon Go' game and lenses of the 'Snapchat' app. Recently, AR has created tremendous market opportunities in the digital world by overlaying real-world user perception. The major rationale of this success is the rapid developments of AR supporting devices and positive user satisfaction feedback.

A traditional AR device is a combination of a head-mounted device and wearable glass devices that provide a virtual world experience to its users. However, in recent years, many multinational companies, including 'Google Vuzix' and 'Microsoft' started to invest in AR supporting devices, which create even better-quality devices that have more advanced human-computer interaction abilities. Specific real-world applications have adopted this fascinating and spreading technology in education, entertainment, healthcare and many other industries to develop applications by providing a high-level user experience in real time.

The AR is misconceptions as just a visual experience, yet visual AR may not be sufficient to provide full user satisfying results since it may distract the user from getting real attractions. However, this effect can be commonly seen in a few instances, such as travel and tourism. Thus, audio augmented reality (AAR) could be a better option in such cases. In addition to the visual experience of AR technology, the AAR can provide audio information as an extra experience. The audio experience of AAR is known to be highly transformative and transportive, with no need for augmented visuals in many domains. It can help to illuminate the background sound effects or music to avoid the distractions. In other words, AAR has the ability to guide the user to concentration along a path by pointing out the most important objects that may not clearly be visible. Therefore, soundscapes of AAR-based applications can be highly recommended to the patients who suffer from vision disabilities to have experience of what they cannot see through the bare eyes.

Besides many other applications, the AAR technology can be integrated into vertebrate vocalisation to enhance ecotourism and wildlife tourism, which is the booming income alternative of many developing countries. Apart from tourism, many other areas may require AAR technology to obtain useful information from vertebrate vocalisation. Some of these areas are shown in Fig. 3.3.

Development of ecotourism and wildlife tourism can conserve the environment and ensure the sustainability of the local people, along with interpretation and education of local wildlife. The AAR-based applications could be a better option to cater to ecotourists as a guide system with a location tracker. This location tracker can be simply built using the vocalisations of vertebrates. As shown in Fig. 3.4, the vertebrate vocal sounds can be integrated with AAR-based applications to allow

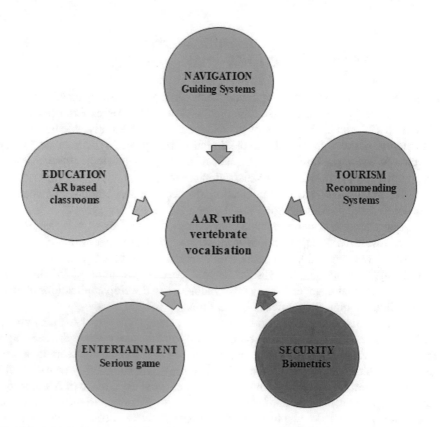

Fig. 3.3 AAR technology with vertebrate vocalisation applicable domains

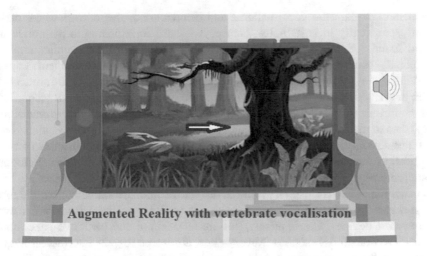

Fig. 3.4 AAR integration for navigation systems with vertebrate vocalisation

visitors to find the location of specific vertebrates easily. Hence, this will be more beneficial to be used in a dense forest environment because capturing the location of small animals and birds is considered as a highly challenging and risky task.

The AAR-based navigation solutions will be helpful to ecotourists if they may visit a place for the first time. Hence, they may face difficulties to navigate from one place to another while exploring wildlife. Apart from navigating solutions to a specific location, this technology can be used as an informative tour guide which can provide the information of vertebrates, that is, breed, age and gender. This may be implemented as a mobile app that can show the vertebrate information in both verbally and visually as text. However, using a mobile app is very common these days due to its easy access. Therefore, the user can get specific information about any interesting vertebrates during an ongoing animal or bird watching task.

The AAR technology can be used to improve the education system, that is, animal science and environmental studies. This can be started by creating awareness on natural conservation to kids from their childhood and make them get involved in protecting the environment for a better future. However, the kids in the twenty-first century are attracted and familiar with modern technologies. Therefore, combining nature-related education with these technologies may provide a positive impact on increasing the learning ability of kids. Moreover, the vocalisation of vertebrates can be used to develop entertainment activities such as serious games to educate the kids about all animals to make them aware of wildlife and conservation, especially about the native animals of their own countries. However, it shows that serious games are one of the most exciting tools in the modern education system that can improve the learning abilities of kids [18].

The traditional classroom-based education system is still considered as the most challenging task to evaluate the understanding of the students about the subject, especially in theory subjects such as zoology. However, it may be difficult for the teachers to explain the vocal patterns and behaviours of the animals at the classroom, specifically in vertebrates. Hence, it is hard for students to memorise the vocal sounds of the animals. Therefore, AAR technology-based applications for vertebrate vocalisation can be integrated with the subject to make students engage well in the classroom. This can make the subject more interactive to the students to keep their interest in the subject and understand its theories better. AAR technology can be further elaborated by the interaction and discussion with three-dimensional (3D) data to make a smart classroom in the near future.

Consideringly, the AAR applications are developing rapidly, but most of these applications are related to human vocalisation. However, lack of other vertebrate vocalisation-based AAR applications is still pending. Almost all current applications use the voice as a primary input to deliver the information. However, the foremost reason for selecting voice as the primary input is because it is adaptable even in low-cost devices. Consequently, gesture recognition and track eye gaze inputs demand advanced AR devices. Nevertheless, compared to gaze interfaces and gesture input, the voice inputs provide high responsiveness and accurate results, but voice-based input suffers a lot from spoofing attacks. Therefore, it requires more attention to security-based research that is related to vocalisations [19].

3.4.4 Virtual Reality (VR)

Virtual reality (VR) offers a real-world immersion experience through the use of VR devices such as Oculus Rift, HTC Vive or Google Cardboard. To make this technology widely affordable and available, several organisations and researchers are working on this domain. Similar to AR technology, most people often concentrate on the visual aspect of VR technology. The function of the audio component is vital for making this technology more efficient. If the images are balanced with what the user hears, the simulated experience will break apart. In VR audio, various audio signals will be fed to each ear to provide the impression of a 3D sound environment, which can be theoretically called binaural recording/beats. Users will experience travelling or being transported into an imaginary world using this technology once VR headphones have been placed on. For example, visiting the jungle is a dangerous activity for youngsters, seniors and people with disabilities, yet they may wish to experience this virtual environment. However, by recreating the soundscape of the forest with the aid of vertebrate vocalisations, VR-based applications may provide them with a high-quality experience. Nevertheless, this device uses location and orientation sensors to synthesise the site's soundscape and play it to the user through headphones [20]. Therefore, the youngsters, seniors and disabled people can experience the forest ecosystem through VR technology at their own residence. This technology can be used for recreating zoos, animal centuries, bird parks, farms as well as aquariums as illustrated in Fig. 3.5. It can help children to learn about wildlife in an informative and engaging way.

Fig. 3.5 Virtual reality for a forest with vertebrate vocalisations

In addition, previous unforgettable memories can be recreated using VR technology. People prefer to capture and store their precious moments of existence in the form of video or audio recordings because of the advancement of hardware and storage devices. Using these recordings, VR can recreate an environment which is identical to any previous incident. Through this technology, an application can be developed for the pet owners to receive a chance to play with their deceased pets. This can be further developed and extended for people who want to play with animals, but experience fears in real life. Hence, people can play and spend their time not only with pets but also with wild animals through this technology. Besides, VR can be used to recreate extinct creatures and globally endangered species to illustrate their historical evolution, habits and lifestyle to future generations.

3.4.5 Cognitive Technologies

Cognitive systems try to mimic humans by adopting reasoning and processing abilities. The fundamental principle of the cognitive system operates by using probabilistic analytics to evaluate data and inferences. In this digital age, data is enormous in volume and increasing rapidly. Therefore, self-learning and constant reprogramming of cognitive technology can help to manage these high volumes of data [21].

The overall efficient navigation of the modern age of computing in a flood of unstructured data can be called cognitive computing. Even these systems are capable of recognising human language subtleties, idioms, idiosyncrasies and nuances. The basic concept of cognitive computing is shown in Fig. 3.6. The concept behind cognitive systems is very remarkable because it attempts to extract the best of both humans and computers to provide solutions in a more extensive, quicker and accurate way.

Researchers suggest that human health monitoring initiatives could accelerate the growth of the health sector, using cognitive technology with big data. There is evidence, especially when considering humans, that voice disorders can be identified by analysing human vocalisation [22, 23]. However, voice disorder analysis is still used as a passive tool in which clinicians obtain and interpret vocalisation data to identify the voice-related pathologies. However, once the patient has recovered, then their data may be eradicated. It would be perfect for healthcare practitioners to make decisions within a short span of time if this form of data analytics could be applied in real time. Similarly, individuals may also know about their health through both self-assessments and expert consultation. However, implementing such a system in real time will still require massive amounts of data. Therefore, it can have significant advantages over traditional systems by combining cognitive technology with healthcare-related applications. Nevertheless, the key points that researchers shall concentrate more on during the development of such applications are ethical and confidential concerns. Furthermore, the vocal data can be stored in a cloud for future use such that a broad range of strategies can be invented to enhance healthcare applications using cognitive systems, yet this shall require the patient's consent.

Researchers discovered new ecological insights with cognitive technological development [24, 25] especially vertebrate vocalisation related applications that need adequate content and reasoning ability. Vertebrates are diverse, so it is a hectic task to store, maintain, and process the collected vocal data. Therefore, a combination of cloud computing, big data and cognitive technologies can bring a new lease of life to this. In addition, many species have fallen into the endangered species category due to human-made actions and climate change. The self-learn and storage

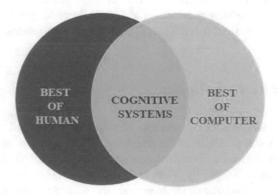

Fig. 3.6 Basic concept of cognitive systems

capability of cognitive computing can be used as a solution to this problem, such that the future generation can learn about the extinct creatures and their behaviour.

3.4.6 Robotic Process Automation (RPA)

Robotic process automation (RPA) is an advanced technology that integrates many engineering disciplines such as electronic engineering, mechanical engineering, automation, network communication engineering and many more [26]. RPA has the ability to build intelligent substances by combining both hardware and software. Many applications use RPA technology to enhance better outcomes, especially in large-scale industries and medical fields. There are plenty of automated and semi-automated instruments that are widely used in these applications. Hence, these inventions require less human interactions and supervision.

In recent years, the world has given a huge amount of attention to autonomous robots. These robots are capable of making their own decisions and doing their tasks as scheduled. These types of robots are mainly used in space investigations, yet many researchers are still in trials to try these robots in the medical field. However, medical robots are now available in the current society. The robots are proficient at performing some surgeries as well as drug delivery. Nevertheless, these robots may be worthwhile to be used in vocal recognition, specifically to act to the commands of humans. However, it is important to recognise the exact vocal sound of the commander to avoid the misuse of the robots. Except for human to robot communication, RPA technology can integrate to build robots to train domestic animals, especially the human-friendly vertebrates such as dogs and cats. If the robot can mimic the owner's voice, then their pets will get trained fast to their owner's commands, yet this robot can provide more benefits by implementing few more necessary functionalities, that is, train to greet the owner and to express the feelings like hunger and thirst, exercising and many other physical activities. This robot can be useful to communicate with the wild vertebrates, specifically for their essential medical treatments. However, a robot can still survive in a dense jungle with the attacks of the aggressive wild animals.

The RPA technology can be used to create automatic machines as an altar solution for many industrial based problems, particularly as a highly accurate and low-cost solution. Industrial agriculture is one of the known areas that is frequently attacked by many wild animals and birds, especially vertebrates. Monkeys, birds and elephants are a few types of vertebrates that attack crops such as paddy fields, fruit and vegetable farms; similarly, they can attack humans. However, an automatic wild vertebrate vocal detection machine can be used as a protection tool. Thus, this tool will not harm both human and wild vertebrates. It is proved that vertebrates make different types of vocal calls to express their feelings, such as hunger, anger, fear and happiness. These calls can be input to the RPA technology to create an automatic repeller to chase away these harmful wild animals and birds.

Moreover, automatic repeller can be used to avoid the predator from animal farms. The predators can cause severe damages to animal farms and destroy the properties. The predators such as hawk, wolf, fox, weasel, coyote, bobcat, bear and snake are the most dangerous predators that can be found in many animal farming areas. However, the number of predators can be increased as per the area. These predators are the most crucial problem that all the farm owners may have. Hence, some of them take brutal actions to address this issue. However, most of the countries have implemented federal laws to protect the predatory animals from humans. Therefore, a vocalisation analyser with the integration of RPA technology can assure a practical solution for this matter. Nevertheless, the predators may keep their vocalisation low during the hunt, but it is certain that the prey, that is, pigs, chicken and other farm animals, would make louder vocal sounds. These vocal sounds can be used to predict the danger and alarm the farm owner; else an automatic security system can be created to activate loud sounds or a human-size robot to work as a repeller to chase away the predators.

3.5 Conclusion Final Remarks

The technological developments have opened more multidisciplinary opportunities worldwide, especially in the medical and engineering fields. Apart from many other opportunities, the audio signal processing takes a significant role because it can provide a wider range of applications with the integration of the trending technologies such as artificial intelligence (AI), augmented reality and virtual reality (ARVR), cognitive computing, internet of things (IoT), intelligent apps (I-Apps) and robotic process automation (RPA). There are many technical integration needs for vertebrate vocalization-based applications, that is, monitoring species biodiversity, conserving endangered species, tracking the environmental pollution, precision livestock farming, frightening tools and biology-based studies. These applications may help to recognise the harm caused by human activity to the natural environment and offers excellent strategies for understanding the biodiversity of species and climate change in the world.

Technology interaction can provide a lot more openings for many vertebrate vocalisation related applications. However, it is important to have proper data collection to achieve higher results from the trending technologies. Thus, the citizen science project is a suitable concept to gather accurate vertebrate vocalisation data using digital platforms. A comprehensive database may use to extract the best feature selections to differentiate the vocal patterns and uniqueness of the creatures. However, technologies such as AI, IoT, augmented reality, virtual reality, cognitive technologies and robotic process automation work well with the best acoustic features.

As discussed in this chapter, there is a certain number of vertebrate vocalisation related applications that can be newly invented via trending technologies. In addition, this chapter provided some suggestions to improve a few existing vertebrate vocalisation related applications via trending technologies. These types of applications may be cooperative to various field of studies to get extreme benefits, especially in medical, industrial, agricultural, scientific research works and many more to improve their findings and innovations. The integration of trending technologies with vertebrate vocalisation is guaranteed to offer many advantages to the world in the future to preserve the ecosystem and its biodiversity. Besides, it will work as a protection shield to conserve the endangered creatures. The advancements of the technologies such as augmented reality and virtual reality can be beneficial to the future generation to learn about the creatures that have been disappeared from the earth. Similarly, the technologies such as AI and IoT can enhance more advantages in many multidisciplinary vertebrate vocalisation related applications, whilst technologies like cognitive computing can provide a platform for handling and storing useful information for research studies to expedite more opportunities and applications in future to achieve better outcomes.

References

1. R.N. Lewis, L.J. Williams, R.T. Gilman, The uses and implications of avian vocalizations for conservation planning. Conserv. Biol. **00**(0), 1–14 (2020)
2. D. Teixeira, M. Maron, B.J. Rensburg, Bioacoustic monitoring of animal vocal behavior for conservation. Conserv. Sci. Pract. **1**(8), 1–15 (2019)
3. B.P.L. Chan, C.F. Mak, J.H. Yang, X.Y. Huang, Population, distribution, vocalization and conservation of the gaoligong hoolock gibbon (Hoolock tianxing) in the Tengchong section of the gaoligongshan national nature reserve, China. Primate Conserv. **31**(1), 107–113 (2017)
4. S. Neethirajan, Recent advances in wearable sensors for animal health management. Sens. Bio Sens. Res. **12**, 15–29 (2017)
5. W. Penar, A. Magiera, C. Klocek, Applications of bioacoustics in animal ecology. Ecol. Complex. **43**(May) (2020)
6. J.C. Bishop, G. Falzon, M. Trotter, P. Kwan, P.D. Meek, Livestock vocalisation classification in farm soundscapes. Comput. Electron. Agric. **162**(April), 531–542 (2019)
7. C.Y.H.G.S. Qiaowei, Detection of laying hens vocalization based on power spectral density. Trans. Chinese Soc. Agric. Mach (2015)
8. H. Dutta, Insights into the impacts of four current environmental problems on flying birds. Energy, Ecol. Environ. **2**(5), 329–349 (2017)
9. J.M. Gilsdorf, S.E. Hygnstrom, K.C. VerCauteren, Use of frightening devices in wildlife damage management. Integr. Pest Manag. Rev. **7**(1), 29–45 (2002)
10. D. Capela et al., Adult male mice exposure to nonylphenol alters courtship vocalizations and mating. Sci. Rep. **8**(1), 1–14 (2018)
11. J.L. Cappadonna, M.F.B.M. Brereton, D.M.W.D.M. Watson, P.R.P. Roe, Calls from the wild: Engaging citizen scientist with animal sounds. DIS '16 Companion Proc. 2016 ACM Conf. Companion Publ. Des. Interact. Syst., 157–160 (2016)
12. D. Stowell, M.D. Wood, H. Pamuła, Y. Stylianou, H. Glotin, Automatic acoustic detection of birds through deep learning: The first bird audio detection challenge. Methods Ecol. Evol. **10**(3), 368–380 (2019)
13. N. Priyadarshani, S. Marsland, I. Castro, Automated birdsong recognition in complex acoustic environments: A review. J. Avian Biol. **49**(5), 1–27 (2018)
14. L. Pozzi, M. Gamba, C. Giacoma, The use of artificial neural networks to classify primate vocalizations: A pilot study on black lemurs. Am. J. Primatol. **72**(4), 337–348 (2010)

15. Scientists create AI that can convert dog bark into human language. [Online]. Available: https://www.indiatoday.in/education-today/gk-current-affairs/story/ai-instrument-converts-dog-bark-into-human-language-1145717-2018-01-15
16. Decoding Animal Talk: How AI May Help Us Speak with Animals. [Online]. Available: https://interestingengineering.com/decoding-animal-talk-how-ai-may-help-us-speak-with-animals
17. A.V. Feng Xia, L.T. Yang, L. Wang, Internet of things. Int. J. Commun. Syst. **25**(5), 1101–1102 (2012)
18. I. Lahbari, H. Alami, K.A. Zidani, *Towards a Passages Extraction Method*, vol 1, no. 1 (Springer, 2020)
19. J. Shang, J. Wu, Enabling secure voice input on augmented reality headsets using internal body voice. Annu. IEEE Commun. Soc. Conf. Sensor, Mesh Ad Hoc Commun. Networks Work., 1–9 (2019, 2019)
20. M. Sikora, M. Russo, J. Derek, A. Jurčević, Soundscape of an archaeological site recreated with audio augmented reality. ACM Trans. Multimed. Comput. Commun. Appl. **14**(3) (2018)
21. D.S. Modha, R. Ananthanarayanan, S.K. Esser, A. Ndirango, A.J. Sherbondy, R. Singh, Cognitive computing. Commun. ACM **54**(8), 62–71 (Aug. 2011)
22. S. Hegde, S. Shetty, S. Rai, T. Dodderi, A survey on machine learning approaches for automatic detection of voice disorders. J. Voice **33**(6), 947.e11–947.e33 (2019)
23. L. Lopes, V. Vieira, M. Behlau, Performance of different acoustic measures to discriminate individuals with and without voice disorders. J. Voice (2020)
24. S.C. Van Hedger, H.C. Nusbaum, L. Clohisy, S.M. Jaeggi, M. Buschkuehl, M.G. Berman, Of cricket chirps and car horns: The effect of nature sounds on cognitive performance. Psychon. Bull. Rev. **26**(2), 522–530 (Apr. 2019)
25. L.M.C. Xu, H. Li, H. Bo, Speech emotion recognition using multi-granularity feature fusion through auditory cognitive mechanism. Int. Conf. Cogn. Comput. **11518** (2019)
26. C.G. Luca Pozzi, M. Gamba, The future digital work force: Robotic process automation (RPA). **16**(9) (2019)

Index

Printed in the United States
by Baker & Taylor Publisher Services